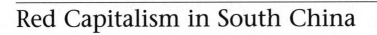

Red Capitalism in South China

Urbanization in Asia

This series focuses on the massive social, economic, cultural, and political transformation taking place as southern Asian countries develop vast urban centres. The future of this highly populated region is an urban one, and the majority of its people will inhabit cities by 2020.

T.G. McGee, professor in the Department of Geography at the University of British Columbia and director of the Institute of Asian Research, is the general editor.

T.G. McGee and Ira M. Robinson, editors, *The Mega-Urban Regions of Southeast Asia*

George C.S. Lin, *Red Capitalism in South China: Growth and Development of the Pearl River Delta*

George C.S. Lin

Red Capitalism in South China
Growth and Development of the
Pearl River Delta

UBCPress / Vancouver

Printed in Canada on acid-free paper ∞

ISBN 0-7748-0616-8

Canadian Cataloguing in Publication Data

Lin, George C.S. (George Chu-sheng), 1962-
  Red capitalism in South China

  (Urbanization in Asia, ISSN 1196-8583; 2)
  Includes bibliographical references and index.
  ISBN 0-7748-0616-8

  1. Chu River Delta (China) – Economic conditions. 2. Urbanization –
China – Chu River Delta. 3. Capitalism – China. I. Title. II. Series.
HC428.C498L56 1997   330.951'27   C97-910350-9

This book has been published with a grant from the Social Science Federation of Canada, using funds provided by the Social Sciences and Humanities Research Council of Canada, and with a grant from the Chiang Ching-kuo Foundation.

UBC Press gratefully acknowledges the ongoing support to its publishing program from the Canada Council for the Arts, the British Columbia Arts Council, and the Department of Canadian Heritage of the Government of Canada.

UBC Press
University of British Columbia
6344 Memorial Road
Vancouver, BC V6T 1Z2
(604) 822-5959
Fax: 1-800-668-0821
E-mail: orders@ubcpress.ubc.ca
http://www.ubcpress.ubc.ca

Contents

Tables, Maps, and Figures

Maps

Figures

Preface

At midnight 30 June 1997, thousands of international dignitaries, journalists, and news reporters swarmed into the newly completed extension of the Hong Kong Convention and Exhibition Centre to witness what may be the last major historical event of this century. After being ceded to Britain in 1842, Hong Kong Island, together with Kowloon and the leased New Territories, was returned to China's sovereignty. As Beijing insisted, the transition ceremony proceeded in a quiet, brief, and low-key manner. Large-scale celebration took place only after the departure of Chris Patten, the last governor of the British colony. While millions of Chinese are celebrating the return of a 'goose' that supposedly will continue to lay 'golden eggs,' few realize that the new government of Hong Kong and that of Beijing are faced with tough political, economic, and geographical challenges.

Politically, the Beijing government has already confronted a dilemma. To maintain economic prosperity and social stability, Chinese leaders are compelled to honour their promise to allow 'Hong Kong people to govern Hong Kong.' But to let a well-established and Westernized civil society go on its own is a risk that Beijing is simply unwilling to take. The active and sometimes provocative participation of Hong Kong civilians in national and international political affairs – such as the dispute surrounding the sovereignty of the Diaoyu Islands – has already troubled, if not disturbed, the government in Beijing. Another concern is the possible northward spread of the spirit of freedom and democracy from this formerly capitalist territory, which may threaten social and political stability on the mainland.

Economically, the Beijing government has informally suggested that it would neither take a penny out of nor invest a penny in Hong Kong. However, this hands-off approach is hard to maintain. Given that economic exchange between Hong Kong and the mainland has been conducted primarily on a local and individual basis that the central government can hardly control, it is meaningless for that government to make promises about investment and revenue collection. To regulate an economy that operates completely according to the fluctuation of the world market will be a new lesson for the Communist planners. To transform an economic infrastructure that remains deeply influenced by the British will also be a formidable task for the Chinese. Furthermore, the government in Beijing has to hope that the economy of Hong Kong will run smoothly for some years to come, as the whole world is watching closely and is ready to point a finger at Beijing if anything goes wrong.

Geographically, both the Hong Kong and the Beijing governments will have to respond to the emergence of some new spatial relationships. The Shenzhen Special Economic Zone, for instance, was originally established as an export-processing zone acting as a buffer between the mainland and the British colony. Now that Hong Kong is officially part of China, what will Shenzhen's new function be? The relocation of Shenzhen's railway terminal to Hong Kong is simply the beginning of a new economic and geographical relationship between the two places. What about the relationships between Hong Kong and Guangzhou, Shanghai, and Beijing? Will Guangzhou continue to function as the economic centre of the Province of Guangdong? Will Shanghai become a direct economic rival of Hong Kong, now that they are parts of the same country? What will be the new function of Beijing? Given the apparent economic edge Hong Kong has over Beijing, will the capital continue to control the national economy? Or will China be governed by two national capitals in the years to come – a political capital in the north and an economic one in the south?

The historic event that took place in Hong Kong has provoked numerous optimistic and pessimistic speculations about the future of the returned territory. Ultimately, the future of both Hong Kong and the mainland will depend on how well the Chinese government takes on the new challenges identified here. But one thing is certain: Hong Kong and the mainland are now tied together more closely than ever before and this link will greatly facilitate the process of economic and spatial transformation that has been under way in China since the open door policy was implemented in 1979.

This book documents and explains how a regional economy in southern China was transformed after being integrated with Hong Kong. As Hong Kong will extend its influence to the entire mainland, this case study of the southern frontier illustrates what may occur in other parts of the country as a result of intensified economic interaction with Hong Kong and the Western capitalist world.

Acknowledgments

Many people have provided assistance for this book. I am especially grateful to Terry McGee, Laurence Ma, and Graham Johnson for their insightful comments and constant support. Their meticulous reading of several early drafts and the critical questions they raised helped me to clarify many ideas introduced in the book.

During the long course of writing and research, I benefited from the intellectual stimulation and wonderful friendship of the UBC Asian research community, particularly from the wisdom and encouragement of Edgar Wickberg, Diana Lary, Samuel Ho, Prod Laquian, Thomas Hutton, David Edgington, Geoffrey Hainsworth, Glen Peterson, Nina Halpern, Yiou-tien Hsing, and Michael Leaf.

I owe special debts to Peter Foggin, Robert North, Trevor Barnes, Victor Sit, Ron Skeldon, Alan Smart, Josephine Smart, Claude Comtois, Francis Yee, Mark Yaoling Wang, Andrew Marton, and two anonymous reviewers for their critical comments and valuable suggestions which have enabled me to improve several early versions of the manuscript.

In various stages of my research, I enjoyed and benefited from scholarly exchange with members of the China Specialty Group of the Association of American Geographers, including Clifton Pannell, C.P. Lo, Jack Williams, Mei-ling Hsu, K.C. Tan, Gregory Veeck, Kam Wing Chan, Cindy Fan, and Chi-kin Leung. Needless to say, I take full responsibility for any errors in this book.

I would like to express my deep gratitude to many scholars in Hong Kong and China for helping me with the fieldwork, particularly to Yue-man Yeung, Anthony Yeh, Wing Shing Tang, and June Chan in Hong Kong; Nora Lan-hung Chiang in Taiwan; Xu Xueqiang, Zheng Tianxiang, Liu Qi, and Chen Lie of Zhongshan University in Guangzhou, and Hu Zhaoliang and Zhou Yixing of Peking University in Beijing.

My appreciation also goes to many supportive colleagues at the

University of Hong Kong, especially to C.Y. Jim, Bill Kyle, P.C. Lai, Yok-shiu Lee, Stewart Richards, James Wang, Youngyuan Yin, and David Zhang. Our cartographers Wong Tingbor and Chen Toytak kindly assisted with map preparation.

The research and writing of this book was funded by the International Development Research Centre of Canada (Grant #91-1025-11), Social Sciences and Humanities Research Council of Canada (Grant #752-92-0925), Social Science Federation of Canada, Chiang Ching-kuo Foundation, and Hui Oi Chow Urbanization Trust Fund. Their generous financial support is gratefully acknowledged. Special thanks are due to Peter Milroy, Jean Wilson, and Randy Schmidt for their encouragement, understanding, and tolerance, which have helped turn the revision of the manuscript from a tedious exercise into a cheerful and enjoyable experience.

Completion of this book would not have been possible without the incredibly strong support of my family. At the height of writing, my wife, Stacey, gave birth to our first son, Jimmy, who has brought to us much joy, love, and inspiration. To Stacey who was left alone for many days and nights bearing and later on looking after a new baby, and to Jimmy whom I held far less often than a new father should, I dedicate this book.

Red Capitalism in South China

1
Introduction

It doesn't matter if a cat is black or white, so long as it catches mice.
– Deng Xiaoping

The mountains are high, the emperor is far away.
– proverb in South China

For scholars studying China, it is always fascinating to see that in this populous country with its extensive history of human civilization, many ideas about how things operate are neatly summarized in popular sayings, proverbs, or idioms. The sayings quoted above are two good examples. Deng Xiaoping had been China's de facto supreme leader since 1978. His words characterize the pragmatic attitude that tolerates a different approach to economic development, regardless of its socialist or capitalist nature, as long as it improves people's lives. The second quotation describes a remote political and geographical environment in southern China where local people have considerable freedom and flexibility to develop a market economy or practise capitalism without having to comply with the rigid rules of socialism laid down by the 'emperor far away' in Beijing.

A major argument I make in this book is that much economic development in China since the reforms has been facilitated primarily by Deng's pragmatism, which allows a market mechanism to grow within a socialist territory. This new pragmatic attitude, combined with some specific local conditions and the growing influence of global capitalist forces, has enabled incipient capitalism to develop in a few selected regions such as southern China and the eastern coast. Moreover, economic liberalization has brought profound changes to the developmental landscape, giving rise to the rapid expansion of new production space in South China, small towns, and the vast countryside.

In the contemporary history of global development, the 1980s was probably one of the most significant decades marking the dramatic collapse of socialism, the end of the Cold War, and the fundamental restructuring of the political economy at both global and regional scales. After competing intensely with its formidable capitalist rival for over half a century, the

Soviet-led socialist empire finally met its doom. Its 'big bang' self-disinte-gration fatally discredited the Marxist prescription of socialism as a supe-rior alternative to capitalism and broke the dream of Communism as the glorious destiny of humankind.

The demise of socialism occurred against a background of the global expansion of the capitalist economy. While countries of the disintegrated socialist bloc are plagued with severe recessions, domestic turmoil, and even civil wars, capitalist economies in North America and Western Europe continue to function well, after adjusting to the ascendance of a new world order. Not only are the capitalist economies capable of making self-adjustments and self-improvements, but they can also further extend the orbit of capital accumulation from national to continental and global scales.

As the tide of capitalism swept the entire world, scholars and develop-ment experts started to re-evaluate existing concepts and models of global development. Important questions have been raised concerning the relevance of Marxist theory to the changing reality. Effort has been made to develop a new paradigm of 'post-Marxism' (Corbridge 1989; Cowen and Shenton 1996; Schuurman 1993). Much attention has been directed to the growth dynamics of the capitalist economies, particularly those in North America and Western Europe, where an economic transi-tion from the traditional stage of Fordist mass assembly to a new era of post-Fordist flexible specialization is said to have taken place (Harvey 1989; Scott 1988, 1992; Massey 1984). In contrast, the transformation of the spatial economy under the seemingly dying socialist mode of pro-duction has tended to be forgotten and relegated to the periphery of development inquiry.

In recent years, however, growing evidence suggests that an abrupt and complete destruction of the socialist system may not be the only form of change for all socialist nations, and that gradual economic transition in countries such as China and Vietnam could generate impressive growth and development. In China since 1978, a series of new economic policies and reform programs has been instituted to stimulate local and individ-ual enthusiasm for production, to attract foreign capital and advanced technology, and to develop a 'socialist market economy with Chinese characteristics.' The result of these economic reforms has been profound structural and spatial change that is no less significant than what has occurred in other world regions. China's double-digit economic growth since the 1980s has few parallels among other nations. Its rapidly expanding national economy is now the world's third largest. An unprecedented experiment affecting one-fifth of humankind and shap-ing a vast country of continental scale is in the making.

The South China Case

This book documents recent economic and spatial development in China since the reforms, using the Pearl River Delta in South China as a case study. Historical and international experience suggests that the industrialization of a country or a region would normally result in rapid expansion of cities, especially large cities that act as the centres of manufacturing production, financial exchange, and modernization. This has not been the case in the Pearl River Delta. Despite the fact that the delta region has been undergoing a remarkable process of rapid industrialization, there has been no growing concentration of population or manufacturing facilities in large cities. Industrialization and urbanization have focused in the countryside, particularly in the areas between or around major metropolitan centres. These areas, usually classified as 'rural,' are not far behind large cities in population density and transportation facilities, both of which are essential to the operation of agglomeration economies. They have been able to attract domestic and international investors and grow at a pace faster than the congested large cities, partly because cheap land and labour could be easily obtained, and partly because regulations on land-use conversion and environmental pollution are not strictly enforced.

On the surface, the Chinese pattern of metropolitan development seems to be nothing more than a simple replication of the American experience of suburbanization or the growth of the megalopolis, a process that can be traced back to the 1960s. However, a close examination of the growth dynamics of Chinese metropolitan development reveals that it is fundamentally different from its American counterpart. As well documented by many geographers, notably Jean Gottmann (1961), the process of suburbanization in North America has been essentially a consequence of urban sprawl. Its contributing forces include the restructuring of the urban economy, growth of tertiary and quaternary activities, automobile and highway development, increased urban income, and growing demand for open space and recreation. These forces have all resulted in the relocation of urban residents from the inner city to the suburbs. By comparison, recent metropolitan development in China has been driven more by the spontaneous growth of industrial and commercial production from the grassroots than by forces of urban sprawl or the relocation of population and manufacturing from large cities.

Metropolitan development in the Pearl River Delta since the reforms has been inseparable from the influence of Hong Kong, which has provided capital and manufacturing facilities to the region. The Chinese case is, arguably, not dissimilar to the American experience of city-driven metropolitan growth. The impact of Hong Kong on industrialization and urbanization of the delta region has been indeed highly significant and will likely become stronger after the return of the British colony to Chinese rule. However, this impact should not be overemphasized, obscuring the role played by rural industrialization and commercialization. A similar process of rural-driven metropolitan development has also been taking place in other Chinese regions such as the lower Yangtze and the Shandong Peninsula, where the influence of Hong Kong is not strongly felt (Pannell and Veeck 1991:115; Zhou 1991:99; Yeh and Xu 1996:246). What has made the Pearl River Delta special is that its proximity to Hong Kong has enabled growth of a much greater magnitude than in other regions of the country.

The rapid economic and spatial transformation of the Pearl River Delta has not been a consequence of active or direct involvement by the central state. Instead, it has been made possible by reduced state intervention, which has allowed local initiative and global forces to interact and promote a spontaneous, self-motivated, and self-sustained economic revolution. Historically, South China in general and the Pearl River Delta in particular have never been a focus of industrial investment from the central state. Even after the reforms, this situation by and large has remained unchanged. What has changed is that the central state removed its constraints on local economic development, allowing the two southern provinces of Guangdong and Fujian to move 'one step ahead' of the nation to develop a market economy (Vogel 1989). Essentially, this compromise represents a trade-off that the socialist state has to make: to decentralize economic decision-making in exchange for local incentive and individual enthusiasm. Whatever the rationale, the removal of state constraints on the local economy has allowed the people of the Pearl River Delta to chart a new course of development for the region that can best use existing comparative advantages. Reduced intervention from the central state has also given rise to a new development mechanism of 'local corporatism,' significantly facilitating incipient capitalism in the southern frontier (Oi 1995:1132).

The prevailing perception is that establishing Special Economic Zones and granting special authorities to Guangdong and Fujian provinces represents an important attempt by the reformed central state to develop the South China region as the 'engine of growth' or a 'growth pole' whose development spirit could later spread to other parts of the country (Yang 1990:241-5; Li 1988:63-70; Cao 1990). I argue that this perception is inadequate. To grant the two southern provinces 'special policy' (*teshu zhengce*) is not so much to give them privileges over other regions but to use them as a testing ground or a laboratory for some experiments that are uncertain and risky for the socialist state. The tacit approach essentially allows more freedom to peripheral economic and geographical areas that are not indispensable to the growth of the national economy. It is another illustration of the state's liberal and pragmatic attitude toward local economic affairs.

Contrary to prevailing opinion, the reformed central state does not have a long-term strategy for national and regional development. The fact that the Chinese are 'building a house without a blueprint' is indicated in the Chinese leaders' description of themselves as 'groping for stones to cross the river' (*caize shiban guohe*) or their development strategy as 'to walk a step and watch a step' (*zou yibu kan yibu*) (Vogel 1989:78; Naughton 1995:5). To have no long-term plan is arguably a practical strategy, because it allows Chinese leaders to adjust instantly and adapt to changing circumstances. Whether this practice represents a strategy or not remains an interesting topic for debate. What is certain is that this cautious and tacit approach has enabled the transformation of the socialist economy through a process of evolution rather than a 'big bang' abrupt destruction. It has also resulted in a spatial transformation that favours those geographical areas with better flexibility and adaptability.

Compared with other regions in the country, the Pearl River Delta is distinguished by its leading role in attracting foreign capital investment and promoting export production. With a population of less than 2 percent of the national total, and a land area of less than 1 percent, the delta region in 1990 contributed a disproportionately high 19 percent of all realized foreign direct investment that flowed into China, and produced 17 percent of the nation's export output. The cause and effect relationships of foreign investment in this region are complex, but the growth dynamics and spatial pattern of such investment can be generally understood as shaped by the interaction of global and local forces. As part of the global process of flexible accumulation, the flow of foreign capital into China, particularly into the Pearl River Delta region, has been motivated by such economic incentives as reducing labour costs, exploring new markets, and ultimately increasing profits. To this end, the Chinese experience has lent support to the new geography of production, particularly to the models of flexible specialization and spatial division of labour that are developed on the basis of the rationale or logic of economic transactions (Scott 1988, 1992; Massey 1984; Harvey 1989).

The imperative of global capitalist accumulation is not, however, the only force shaping the growth dynamics and spatial distribution of foreign investment in the Pearl River Delta. When analyzed more closely, the establishment and operation of many manufacturing or business firms in a specific place is often the result of some particular local social relationship, such as kinship ties, interpersonal trust, and connections (*guanxi*) between overseas investors and local cadres (Leung 1993, 1996; Smart 1993, 1995). Such social relationships are not excluded from the sphere of economic transactions. Instead, they provide the most effective and secure channel for global capitalism to take root in socialist soil. The logic of global capital accumulation and the logic of local particularities therefore do not work separately. Nor do they meet head-on. Despite some occasional conflicts, global and local forces have been able to interact in a cooperative manner so that their different interests could eventually be realized. This process whereby global capitalism does not conquer but seeks shelter from local traditions has been one of the important factors leading the socialist economy to transform gradually rather than abruptly. It has significantly promoted the growth of incipient or 'local capitalism' in the delta region (Smart 1995). Geographically, it has contributed to the rapid expansion of local economies in some suburban areas where kinship ties between local people and overseas investors are strong.

Theory and Reality

The Pearl River Delta region has raised some interesting and important theoretical questions about the operating mechanism of urban and regional development in China. For years, scholars studying China tended to view urban and regional development as a process driven primarily by the growth of large cities, by the intervention of a powerful socialist state, and by some external forces, usually capital investment and market demand. Their assumptions about the pivotal role played by large cities, the state, and external forces in national and regional development have been central not only to many theories about Chinese spatial transformation but also to the general wisdom about Third World development. These assumptions may need to be re-evaluated in light of the rapid changes taking place in post-reform China, particularly in the South China region.

Although significant disagreements exist about the impact of urban growth on national and regional development, it is generally believed that cities, especially large cities, function as the centres of industrialization and social change. We have been well informed that cities act as the 'centre of change,' simply because of their many comparative advantages over smaller human settlements. Cities are portrayed as the places where economies of scale can develop and operate, where advanced technological innovations originate, and where underdeveloped countries were contacted initially by Western colonial powers. This leading role for cities has been a tenet or axiom in the conventional theory of economic growth and modernization, although its impact was variously interpreted as positive and catalytic, or negative and parasitic (Lin 1994a:8). As China seemingly makes its move to follow the global practice of industrialization and modernization, it is not surprising to see existing urban theory used to explain the Chinese reality. Such a direct application of Western theory to China may not, however, be sensitive to what has actually occurred in the country, particularly to its rapid development since the reforms.

It is true that cities play a leading role in the process of industrialization and urbanization, but such a role is not static over time nor homogeneous across space. Cities act as centres where population and economic activities are concentrated. As the logic of the spatial organization for population and economic activities changes, the role of cities in regional development, and the positions they hold in the settlement hierarchy, will also change. Cities also function and develop within a particular historical, political, and geographical context. Their patterns of growth always vary according to the changing nature of the current political economy.

Before the reforms, Chinese cities, especially large cities, were able to dominate the developmental landscape of the country, because they fit well into the political and economic agenda of the socialist state. The adoption of a development strategy aiming at high-growth rates and oriented toward heavy industry entailed a high concentration of manufacturing activities in large cities. A centrally planned economy organized by vertical linkages required large cities to integrate all settlements effectively, and to transmit decisions from the top to the bottom of the settlement hierarchy efficiently. The socialist ideology of anti-commercialism undermined the economic base of many marketing centres in the countryside and reinforced the primacy of large cities in the settlement hierarchy. As well, the unwillingness and inability of the socialist state to accommodate urban expansion precluded the upgrading of small cities and elevated large cities to a position well above small settlements.

This peculiar political and economic situation changed profoundly after economic reforms were instituted in 1978. The goal of development is no longer to seek the high-growth rates of city-based industrial production, but to raise income and profits for the general population, especially those who live in the countryside. Decentralization of decision-making has given local people the incentive and flexibility to engage in manufacturing, commercial, and other non-agricultural activities. Introduction of the market mechanism has helped an industrialized and urbanized economy to grow spontaneously outside of the central plan and away from the large cities, as has been well demonstrated in the Pearl River Delta. These changes have created a new political and economic context where cities have a new role to play. Given this situation, can we still perceive Chinese cities as the centres of industrialization and urbanization? Is it tenable to view large cities as the most dynamic centres of change? Does rapid industrialization necessarily lead to the growing concentration of manufacturing activities and population in large cities? What is the new relationship between large cities and small towns, and between cities and countryside? What are the similarities and differences between the Chinese experience of spatial transformation and the process of flexible specialization that has already occurred in

North America and Western Europe? These questions all demand further investigation.

The changing nature of the Chinese socialist state is another important issue that requires reassessment in light of the locally driven development of the Pearl River Delta region. Traditionally, national and regional development under state socialism has been perceived as a process effectively manipulated by a powerful socialist state that monopolizes all economic affairs ranging from production to circulation to redistribution. State intervention was indeed pivotal to the creation and transformation of spatial economies in many socialist nations during the Cold War. Sufficient evidence suggests that socialist China under Mao was no exception. In recent years, however, implementation of liberalized economic policies has significantly shifted decision-making from the central to local governments. This shift has given rise to a new power structure in which local cadres and individuals have an active role in charting regional development. The new pattern of locally driven development has been especially noticeable in regions such as South China where the direct intervention of the central state has been substantially reduced.

Local initiatives are, of course, essential to the use of regional comparative advantages and promotion of genuine economic development. However, because of the different interests of the central state and the local governments, the effect of a local initiative may not always be desirable to the central state. The result has been a complicated political relationship between the central state and the local governments, and among governments of different regions. In general, such a relationship is cooperative and mutually accommodating, but it is not without tension and conflict. Given the increasing role played by local initiative and the growing complexity of the central-local relationship, is it still sensible to view Chinese economic development as a process effectively manipulated or monopolized by a uniform and powerful socialist state? Should a distinction be made between central state and local state, and between nation state and region state? Is it necessary to differentiate the power of the state to formulate policies from the genuine ability of the state to implement such policies? What are the new roles played by the central state and the local state, and how do they interact? More importantly, how has this restructured political system transformed the Chinese spatial economy?[1]

Finally, the relationship between external and internal forces in shaping the pattern of regional development requires re-evaluation, given the complex growth taking shape in the delta region. In documenting the recent process of spatial transformation, current prevailing theory has described the Chinese practice as an endorsement of the neo-classic school of thought that prescribed an unbalanced or polarized approach to regional development (Yang 1990; Li 1988; Linge and Forbes 1990; Hsu

1991). We have been well informed that the new government under the pragmatic leadership of Deng Xiaoping has opted for efficiency over equity, individual creativity over egalitarianism, and openness to the outside world over self-reliance. Our attention has been drawn to the fact that the Chinese government is purposely and selectively developing Special Economic Zones, Open Cities, and Open Economic Regions, and using the coastal area as the 'engine of growth' or a 'growth pole' whose effects will automatically 'trickle down' or spread out to the less-developed regions in the interior and the countryside. According to this perspective, the growth pole strategy, which was fashionable in the 1950s and 1960s among scholars and planners working on Third World development, is now enjoying a renaissance in post-reform China. In this vision, recent development in China is nothing more than another case of the rationality and applicability of the theory of polarized growth that views regional development as an exogenous process driven by external market demand and capital investment.

The neo-classic interpretation of the Chinese approach to spatial transformation is flawed because of the obvious inconsistency between theory and reality. At least three critical issues have not been clarified. First, the application of the strategy of polarized growth will theoretically result in a growing concentration of population and economic activities in the few geographical centres selected for concentrated investment. The consequence of this 'polarization' or 'backwash effect' will be an increased disparity of output and income between the core and the periphery (Hirschman 1958; Myrdal 1957; Friedmann 1972). This outcome has not been the case in China, according to the many empirical studies that document changing regional inequality in the country since the reforms (Fan 1995; Lo 1990; Wei and Ma 1996; Xu and Li 1990). Scholars are puzzled by the fact that the interprovincial inequality of production capacity and income generation has been reduced, despite the implementation of a strategy that seems to take the form of polarized growth. Some writers have attributed the Chinese 'anomaly' to the role played by the socialist state, which maintains control over investment, production, and distribution. Such a perception has not, however, provided a satisfactory explanation and convincing evidence to show how the socialist state has, on the one hand, concentrated its investment in a few selected centres, and on the other hand, immediately redistributed the benefit of growth from the core to the periphery. Second, the prevailing interpretation of the Chinese experience has been based on the assumption that the socialist state continues to play a pivotal role manipulating growth and redistribution of the spatial economy. Recent studies have increasingly shown that this assumption may no longer be adequate to shed light on the country's complicated development mechanism. A growing number of scholars have revealed that what has been taking place in China since the reforms is a result of state deregulation or disarticulation rather than increased state intervention (Oi 1995; Walder 1995a; Liu 1992; Ma and Fan 1994; Ma and Lin 1993). Finally, the direct theoretical application of the concept of a growth pole has not provided a sensible explanation for the rapid economic growth and spatial transformation of the small towns and the vast countryside. These places have not been selected by the state as growth centres for concentrated investment, but they have demonstrated growth no less, if not greater, than that of the chosen cores. In the Pearl River Delta, for example, industrial and transport development since the reforms has been focused in small towns and the countryside where little investment has been received from the central state (Johnson 1992; Lin and Ma 1994).

The contradictions identified here between theory and reality raise serious questions about interpreting the Chinese pattern of post-reform development as a simple duplication of polarized growth. To what extent has Chinese regional development been driven by external forces, particularly by investment from the state and from foreign sources? Is the process of post-reform development mainly exogenous or endogenous in nature? Does the changing emphasis from equity to efficiency and from isolation to active participation in global development necessarily lead to domination by external forces? How have external forces penetrated into a Chinese region and promoted its economic and spatial transformation? Do they overtake, cooperate, or depend on local indigenous forces? Is the interaction of external and internal forces consistent over time and homogeneous across space? If yes, what is the general pattern of interaction? If no, how does it fluctuate through time and vary from place to place? Furthermore, in what ways have internal and external forces interacted to rearrange the spatial distribution of economic activities, population, and land use?

The answers to these questions will inevitably vary according to personal opinions, individual perspectives, and values. However, they must ultimately rest upon solid investigations of real-world cases. What I present in this book is essentially a detailed case study of recent economic and spatial transformations that have been taking place in one of the fastest-growing regions in post-reform China. The objective is to identify the pattern of economic and spatial changes, to determine the key forces responsible for such changes, and to explore the theoretical implications of those changes in the broad context of regional development. Granted, what has occurred in the highly developed Pearl River Delta region may not be typical of the general situation in the country. However, by examining in detail the process of economic and spatial transformation in a region that has moved 'one step ahead' of the nation, we can develop significant insights into what may occur in other Chinese regions, if they are allowed to follow the development pattern of the delta. Moreover, as other regions of the country are opened to foreign investment, the Pearl River Delta may no longer maintain its leading economic position in the nation and its development experience may gradually lose its uniqueness. If this is the case, then this study of a pioneer region's development will provide important lessons for regions that have just been exposed to similar global market forces.

Terminology

Before we turn to the complex processes and spatial pattern of the recent changes in the Pearl River Delta, several terms used frequently in this book require clarification. The first term is capitalism, which has been used in numerous publications. Despite its wide use, its precise implications remain vague and elusive. To those who follow the neo-classic school of thought, capitalism seems to refer to an economy where the 'free market' or 'free enterprise' operates. To others who endorse Marxism, capitalism represents a distinct 'mode of production,' or a social system that exists between feudalism and socialism. No consensus has been reached on what the term should mean. This lack of consensus is also apparent in the documentation of recent economic transition in China. A growing number of scholars have conceptualized the Chinese experience as 'capitalism with Chinese characteristics' (Vohra 1994; Smart 1995; Redding 1990; Solinger 1989), but very few have spelled out a definition generally acceptable to all.

Because of the lack of consensus, the concept of capitalism used in this book is defined in a case-specific manner. It denotes a new type of production mechanism, institution, and social order that is developing in the Pearl River Delta region. It has been operating most noticeably in at least three distinct spheres. First, a free market economy is growing rapidly outside of any centrally planned system, in which local people have the freedom to decide what and how production is to be carried out. As well, prices can fluctuate according to the changing relationship between supply and demand, and output is redistributed according to individual productivity. This economy is also characterized by the free entry and competition of firms, the mobilization of capital through selling stocks, the conversion of some state-owned industrial assets into shares, successful enterprises taking over failing ones, and the bankruptcy of some hapless plants. A unique Chinese characteristic is that many enterprises in the countryside remain collectively owned, but they are actually operated by local cadres who act as the equivalent of the board of directors. Second, a distinct culture and society is developing, where people are motivated more by profit-seeking and material rewards than by abstract revolutionary propaganda, symbolic rewards, or moral exhortation. Finally, a locally driven process of spatial transformation has evolved, in which production activities can be spatially rearranged according to changes in market demand, land can be transferred or subcontracted, and people can move from place to place for economic pursuits. One down side of this indigenous process of spatial development has been the problem of the unplanned duplication of production facilities, chaotic land use, and environmental degradation.

Although the operation of the capitalist mechanism in the Pearl River Delta region remains under the constraints of Communist rule, the aforementioned features are distinct from those associated with classic state socialism. What I call capitalism here shares some common features with what others have called market socialism (Kornai 1992). However, I have opted for the term 'capitalism' over 'market socialism' primarily out of consideration for two important factors. While China as a whole has been undergoing transition from state socialism to market socialism, there are reasons to believe that South China, particularly the Pearl River Delta, has moved ahead of the nation to experiment with capitalist forces. Also, what has been taking place in South China may not yet warrant the label of classic capitalism, as the development is still in an early stage of growth. However, in the long term, the general trend of this development is a direction that is increasingly more capitalist than socialist.

The areal extent of South China and the Pearl River Delta also requires specific demarcation. Conventionally, China is divided into seven geographical regions, where South China includes the four provinces of Fujian, Guangdong, Guangxi, and Hainan.[2] Only the two provinces of Guangdong and Fujian are discussed in this study of South China. This limited definition is partly because Guangxi and Hainan share little in common economically with the other two provinces despite their southern location, and partly because data on Guangxi and Hainan are not detailed enough to make a systematic comparison. The focus of the region is the Pearl River Delta, which is defined as an area that consists of the provincial capital city of Guangzhou, two Special Economic Zones (Shenzhen and Zhuhai), and twenty-eight other cities and counties. The delta region so defined covers an area of 47,430 square kilometres and has a population of over 20 million.[3]

The concepts of economic growth and economic development have often been used interchangeably in many publications, but they are treated differently in this book. Growth means only an increase in output whereas development refers to a process characterized by improved productivity, increased employment, and higher per capita income. Growth may occur with or without development.

In this book, the term 'state' is used to refer mainly to the central state or central government. In China, however, the levels of state organization range from the central government to those of the province, municipality, county, township, and village. My argument is that the central state's involvement at the local level in the Pearl River Delta region has not been a major factor in its post-reform development. Instead, that development has been characterized by the active direction of local organizations at the county, township, and village level.

Data for this book were gathered through field investigations, interviews, and documentary research conducted in China between 1980 and 1995. For the purposes of comparability and consistency, I focus on the period 1980-90 for the systematic assessment of post-reform economic and spatial development. The historical context before 1980 and events after 1990 are discussed wherever possible. Selecting the 1980s as the period under study has allowed me to use economic data that are measured at constant 1980 prices and at the level of county and city proper. While the output value, population, and land resources may have changed since 1990,[4] the nature of growth or the mechanism of economic and spatial transformation remains more or less the same. Just like an economic assessment focusing on an industrial sector, or a historical investigation based on a particular period, this case study of growth and development of a most dynamic region in its first decade of reforms may shed signifi-

Map 1.1

Geographical division of the People's Republic of China

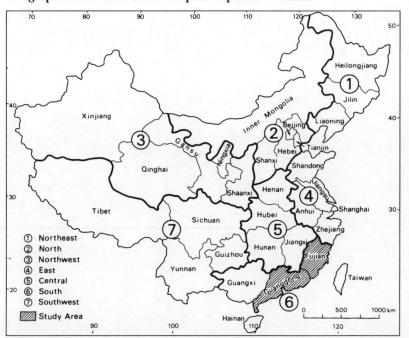

Source: Pannell and Ma 1983:5; Wu 1967:21; Kirkby 1985:146.

cant light on how a regional economy under socialism is transformed after the intrusion of capitalist forces.

Organization of the Book

Over the past four decades, China's national economy has undergone profound structural change. The transformation of the Chinese political economy before and after the 1978 reforms has been well documented (Walder 1995b; Hsu 1991; Nee 1989; Naughton 1995; Oi 1995; Lardy 1978; Huang 1990; Ho 1994). However, with few exceptions (Skinner 1994; Johnson 1992; Solinger 1987; Yang 1990; Chen 1994), the documentation of those political and economic reforms has seldom been related to the process of spatial restructuring. Yet without an investigation of the geographical consequences of political and economic change, any attempt to account for what has been taking place in China must be necessarily limited and incomplete, because no economic policy or development strategy has ever been successfully implemented without being grounded in some specific geographical areas. This investigation begins in Chapter 1.

Chapters 2 to 4 discuss the links between the transformation of the political economy and the restructuring of the developmental landscape

Map 1.2

The Pearl River Delta region, 1990

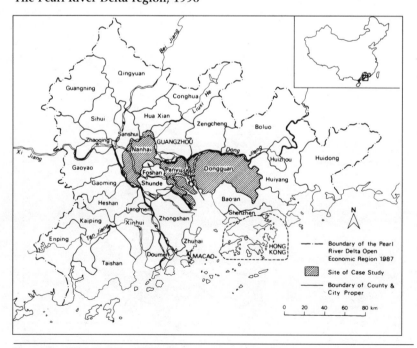

Source: Chu 1996:475; Zheng 1991:4-5.

at the national scale. An important issue highlighted here is the implications of different political ideologies and national development strategies for regional economic growth in South China.

Chapter 5 systematically assesses the dramatic economic and spatial changes that occurred in the Pearl River Delta in the first decade of reforms. Analyses of regional data have clearly revealed a spatial pattern of growth, in which production activities and population become increasingly concentrated in the triangle zone bordered by Guangzhou, Hong Kong, and Macao. There was no excessive expansion of population and production activities in the primate city of Guangzhou, as the conventional wisdom of urban transition might have predicted.

In Chapters 6 through 8, three detailed case studies show how various local and global forces have contributed to the transformation of the regional economy in South China. These studies are carefully chosen according to the suggestions of experienced local Chinese researchers and my own field investigations. Finally, Chapter 9 summarizes the main features of economic and spatial transformation of the Pearl River Delta and discusses their theoretical and practical implications, which includes a model of local-global interaction. The results of empirical research are also related to some current theoretical debates on the mechanism of regional development in the Third World.

Notes

1 A spatial economy is defined as an integrated and organized system of economic activities, population, and land use, which takes place and operates in a particular geographical confine. It could range from the national to regional to local scale.

2 It is customary to divide the country into seven regions according to their different geographical characteristics. They are as follow: Northeast (Liaoning, Jilin, and Heilongjiang), North (Hebei, Shanxi, and Inner Mongolia), Northwest (Gansu, Shaanxi, Xingjiang, Qinghai, and Ningxia), East (Jiangsu, Zhejiang, Shandong, and Anhui), Central (Henan, Hunan, Hubei, and Jiangxi), South (Guangdong, Hainan, Guangxi, and Fujian), and Southwest (Sichuan, Yunnan, Guizhou, and Tibet). See Map 1.1.

3 I have adopted the 1990 statistical definition of the Pearl River Delta (*Zhujiang shanjiaozhou*) region, which includes Guangzhou, Shenzhen, Zhuhai, Foshan, Jiangmen, Huizhou, Zhaoqing, Zhongshan, Dongguan, Shunde, Nanhai, Panyu, Xinhui, Qingyuan, Baoan, Doumen, Zengcheng, Gaoming, Heshan, Taishan, Kaiping, Enping, Sanshui, Huaxian, Conghua, Huiyang, Huidong, Boluo, Gaoyao, Sihui, and Guangning. See Map 1.2. All data are measured at the level of county and city proper, because administrative boundaries at this level were relatively constant during the 1980s. Several counties were promoted to city status (for example, Zhongshan and Dongguan) during this period, but their original county level boundaries remain unchanged.

4 In the Chinese statistics, output value is usually calculated in Chinese yuan. The average official exchange rates between US$1 and Chinese yuan were 2.32 in 1984, 2.94 in 1985, 3.45 in 1986, 3.72 in 1987 and 1988, 3.77 in 1989, and 4.78 in 1990. I use the 1990 official exchange rate (US$1 = 4.78 yuan) in this study. See International Monetary Fund 1991. *International Financial Statistics* 44 (3-4):164. Land area in China is measured by mu, where 1 mu = 0.0667 hectare = 0.1647 acre.

Part 1:
National Context

2
The Operating System of Spatial Transformation

The development of a spatial economy, in a national or a regional context, is a complex phenomenon attributable to many political, social, and economic factors. The effect of political forces in shaping a national economy is especially evident in socialist China, where until recently a powerful Communist regime had monopolized social and economic affairs. To comprehend the issue of spatial development in China, we must go beyond the geographical and economic sphere and instead analyze major political forces, including those operating at the local, national, and global levels. As well, the transformation of a spatial economy is a historical phenomenon. The geographical organization or spatial arrangement of economic activities, population, and land use has constantly changed over time as a result of shifting leadership, ideology, and development strategies. An analysis of changing historical situations is therefore essential to understanding the growth dynamics of spatial transformation.

The next three chapters examine the operating mechanism and growth dynamics of the Chinese spatial economy. The discussion addresses two interrelated issues. First, who are the key players responsible for Chinese spatial economic development? What are the driving forces behind the scene of spatial transformation? Second, how has the Chinese spatial economy evolved since the Communists seized power in 1949? More specifically, how has the developmental landscape changed from the Maoist to Post-Mao era? Included in this discussion are analyses of the implications of changes in the national context for the growth and development in the Pearl River Delta.

The Operating System

In general, the Chinese spatial economy can be seen as a system operated by three key players, namely, the central state, local governments, and global capitalism. Until recently, national economic development has been effectively controlled by a powerful socialist state, which, by launching endless political campaigns and implementing Stalinist-styled command economic policies, manipulated the spatial economy so that any changes would neither jeopardize the grand socialist-communist enterprise nor endanger national integrity and security. Since the 1978 economic reforms, the role of the central state has remained important, and under specific circumstances crucial, as evidenced by the significant impact of drastic policy change immediately following the 1989 Tiananmen incident.

The central state is not, however, the sole player responsible for national and regional development. Since the late 1970s, new reform programs have decentralized much economic decision-making from the central state to local governments, at the same time opening China for foreign investment and international trade. This shift has brought into play new forces emanating from the global and the local spheres. Increasingly, change in the spatial economy has been the complex outcome of the interplay of central state, local governments, and global forces, all of which require scrutiny if we are to understand the operating mechanism of development at both the national and the regional levels.

The Socialist Central State

For most of the years since the Communists took power in 1949, the state followed the Soviet practice of state socialism, developing a centrally controlled system in which the state monopolized nearly all of the nation's political, economic, and social affairs. The role played by the central state in spatial economic development is most visible in three areas – politics, economics, and demographics.

Politically, the socialist state until recently had been occupied with two major tasks: to prevent the Chinese from contamination by capitalism, and to protect the nation from possible attack by hostile international powers. To pursue these goals, the central state adopted some extraordinary development policies. On the one hand, these policies severely constrained urban consumerism and limited rural commercial activities, because consumerism was considered the root of capitalism. On the other hand, they forced capital investment away from the coastal zone toward the interior regions, because that zone was perceived as vulnerable to military attacks from Japan, South Korea, Taiwan, and the United States. This political context, dominated by ideological and military concerns, was extremely unfavourable to economic growth on the eastern and southern coast, not only because their frontier locations were considered 'insecure,' but also because their trading tradition and commercial activities were seen as dangerous 'seeds of capitalism.'

Economically, since the early 1950s, the United Nations' economic sanctions against the Chinese involvement in the Korean War had cut off trade links between China and the Western world. Confronting a hostile international environment, the socialist state endeavoured to develop a self-reliant economy by emphasizing the production of food grain and basic manufactured goods, particularly iron and steel. In the countryside, where the majority of the population lived, grain production was universally viewed as a top priority in the allocation of land resources and manpower. This single-sided economic policy, imposed by the central state, left little room for local initiatives in other specialized commercial activities more suited to local conditions than grain production.

Demographically, the central state allowed the population to grow for most of the 1950s and 1960s, while strictly limiting rural-urban migration. On the one hand, the Maoist vision of people as the most precious labour asset precluded any attempt to curb population growth. On the other hand, a socialist 'economy of shortage' running on a limited amount of capital meant that the state couldn't afford massive rural-urban migration, which would require substantial investment in urban infrastructure and public facilities. To preserve the available capital for industrial development and to ensure social stability, the central state in 1958 introduced a household registration system to ration the supply of foodstuffs and to limit housing and educational services to people in designated urban places. This system effectively restricted rural to urban migration.

The result of state monopoly had been a stagnant national economy operating at a subsistence level. Many people were restricted to a rural life, without any opportunity and incentive to specialize in those economic activities for which they had a comparative advantage or expertise. Although the statistical record for this period shows considerable increase in the production of food grain, iron, and steel, there was in reality no significant improvement in the standard of living. It has been recorded that from 1949 to 1978, grain output increased from 113 million to 305 million metric tons, and steel production from 0.158 million to 31.78 million tons. However, such a remarkable increase in production had been offset by rapid population growth from 541.67 million to 962.59 million. Consequently, the net agricultural output increased on average by only 1.9 percent between 1953 and 1978. Per capita rural income increased slightly from 103 yuan to 113 yuan in the three decades between 1957 and 1977. Per capita grain output dropped from 3.6 kilograms to 2.9 kilograms in the same period (China, State Statistical Bureau 1983:158, 245, 103). As much of the population was restricted to a single-sided agrarian economy, the nation suffered from severe unemployment or underemployment despite the growth of agricultural and manufacturing output. It was estimated that, on the eve of economic reforms in 1978, unemployment ranged from 10 to 29 million in the cities and from 40 to 90 million in the countryside (Yeh 1984:693). If we make a distinction between 'growth,' which refers to the increased absolute amount of output or production value, and 'development,' which means an actual improvement in the standard of living, the economic consequence of the socialist state monopoly before the reforms could be conceptualized as 'growth without development' or 'involutionary growth' (Huang 1990:13).

The death of Mao in 1976 and the subsequent takeover led by Deng Xiaoping ushered in a new pragmatic government that plays a different role in China's national and regional development. Aware of the deficiencies and inefficiency of the previous system of state socialism, this new administration introduced a series of reform programs that shifts much administrative power to the local governments to stimulate local enthusiasm (Jia and Lin 1994). Economic decision-making no longer had to be justified according to the classic writings of Marx, Lenin, and Mao. Local governments were instead encouraged to engage and specialize in those economic activities that can use local strengths to the greatest extent, including trade and commercial production. Many cities and regions in the coastal zone were given special authority to attract foreign investment and acquire modern technological know-how. In the countryside, the state-monopoly network for the procurement and distribution of basic farm products (*tongguo tongxiao*) was abolished and the administrative system of commune-brigade-team was dismantled. Since 1984, control over migration has also been relaxed, and peasants are now allowed to move into nearby towns, provided that they can satisfy their own needs for foodstuffs and accommodation. The implementation of these new policies made state intervention less rigid and significantly promoted national and regional development.

Local Governments

Another operating force with an increasingly important role in shaping the Chinese spatial economy is the development initiative of local governments. Before economic reforms, the local governments and the people of the community, individually or collectively, could hardly do much more than try to satisfy the political and economic demands of the central state. Under the rigid socialist command economy, the central state was the chief economic agent, setting production targets, providing most of the raw materials for production, and distributing major agricultural and industrial outputs. After fulfilling the production targets set by the central state, local governments would then get a certain amount of capital from the state to maintain production facilities, repair or replace obsolete machinery, and undertake urban construction. Such a system deprived local people of their creativity and freedom to do those activities for which they had special skills or comparative geographical advantages.

Since the reforms, a new management system of 'fiscal responsibility' (*caizeng baogan*) has been introduced. Under this system, local governments are allowed to choose their economic activities, provided that they make a lump-sum payment of profits to the central state (Vogel 1989:89; Cheung 1994:213). The remaining profit can be used by local governments for their own purposes. The implementation of this policy greatly inspired local governments to engage in those economic activities that can generate high profits, including trade, commercial agriculture, and consumer-goods manufacturing.

In agriculture, the emphasis has shifted to sideline commodity farming, livestock husbandry, aquatic production, and especially manufacturing, all of which are generally known to be more profitable than traditional paddy rice cultivation. As a result of such profit-seeking local initiatives, agriculture has been increasingly commercialized and the rural economy rapidly industrialized (Byrd and Lin 1990; Ho 1994; Johnson 1992). Since the 1980s, many township and village enterprises have been set up, most of which are operated by local cadres acting as the equivalent of a board of directors, or sometimes more directly as chief executive officers. This development has given rise to a distinctive system of decision-making, operating under but not necessarily following the guidance of the central state. The new system, called 'region-state' by some, 'local state corporatism' by others, or 'feudal-lord economies' by the Chinese, has become a prominent feature that distinguishes spatial economic development before and after the reforms (Ohmae 1993; Oi 1995; Walder 1995a; Jiang 1990; Shen and Dai 1990). In any event, this locally driven developmental mechanism has resulted in unprecedented large-scale rural industrialization. As industrial production expands and its profits increase, the local revenue base has grown and consequently more development funds are available for infrastructure development and urban construction. In turn, this development has significantly increased the pace of rural urbanization. The combined result of commercialization and industrialization has been a process of genuine development, substantially improving the productivity of labour and land (Huang 1990:18).

More importantly, local initiative has played a crucial role in mobilizing capital from various sources to build a modern transport infrastructure system, a step that is generally considered a prerequisite for economic growth and for attracting foreign investment. In recent years in the Pearl River Delta, many cities and counties have set as their top priority building a modern telecommunication system and an efficient road transport network. By building bridges, roads, highways, harbours, ports, and even airports, Pearl River Delta economic planners have managed to overcome the 'friction of distance' and create a transactional environment conducive to economic development. This process of spatial contraction, or what

Harvey calls 'time-space compression' (1989:260), has significant implications for the redistribution of economic activities, population migration, and land-use transformation.

The Influence of Global Capitalism

Although Chinese spatial economic development in the Maoist era was not exempt from the impact of changing international situations, the large-scale penetration of global capitalism into China has been a fairly recent phenomenon. It started in the late 1970s, when the open door policy was introduced, under which foreign investors are offered such preferential treatment as tax concessions and duty-free imports of machinery and equipment. Driven by the incentives of low-priced labour and cheap, unregulated land, a growing number of overseas manufacturers, most of them from Hong Kong, moved into China to set up joint ventures, export-processing plants, or cooperative businesses (Ho and Huenemann 1984:29; Lardy 1987; Yeh and Xu 1996:227).

Geographically, foreign investment has concentrated in the four newly established Special Economic Zones, in the coastal cities that have been granted special authorities to do business with foreign investors, and in the officially designated Open Economic Regions, such as the Pearl River Delta (Yeung and Hu 1992; Edgington 1986; Yee 1992). By creating employment opportunities for the local population, the inflow of foreign investment has helped absorb surplus rural labour, facilitated the industrialization of the rural economy, and contributed to the process of genuine regional development.

The influence of global capitalism is especially evident in South China, particularly in the Pearl River Delta region, where personal connections with overseas Chinese are extensive and economic linkages with Hong Kong are strong. The penetration of global capitalism in South China has strongly relied on some pre-existing social relations such as kinship ties and cultural linkages (Leung 1993; Smart 1993; Smart and Smart 1991). In many cases, global capitalism does not conquer or compete with the local economy. Instead, it becomes subordinate to local conditions and grounded in some local informal processes that better guarantee economic and social interests (Smart 1995). In this regard, the Chinese experience is distinct from those of other 'frontiers of capitalism,' where global capitalism is an external system that goes head to head with the local society. This distinct pattern of global capitalism penetration has important implications not only for the growth of the local economy, but also for the transformation of the regional developmental landscape.

Summary

The roles played by the central state, local governments, and global capitalism are summarized in Figure 2.1. These three key players do not, however, act alone in shaping the Chinese spatial economy. Instead, they frequently interact to change the pace of national economic growth and alter the developmental landscape. For instance, the relaxation of central state control made room for local governments and global forces to play their roles. Local development initiative in promoting commercialized agricultural and rural industrial production would not have succeeded without the permission of the central state. Although the state is not actively involved in this process, it sometimes tries to control the growth of market farming and rural industry when conflicts between central and local interests arise. Similarly, the penetration of global capitalism could only take place after it was adapted and sheltered by some deep-rooted local economic and social relations in which local governments play a crucial part. This distinctive mechanism of local-global interaction, in which global capitalism gradually penetrates into the socialist system through local, informal social relations, is an important factor that explains why the market transition in China has been a gradual process. This transition differs markedly from the sudden big blow or 'big bang' that characterized the transformation of other socialist economies, notably in the former Soviet Union and Eastern Europe.

Moreover, the roles played by the central state, local governments, and global capitalism have changed according to different political and economic situations. Although socialist China is ruled by a single Communist party, significant differences on many issues have long existed among top

Figure 2.1

Operating system for spatial transformation in China

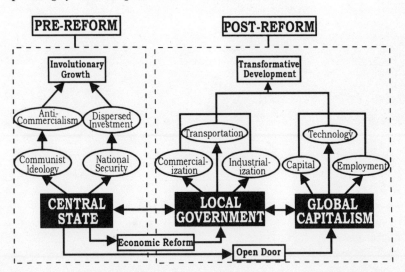

party leaders and government officials, including interpreting classic Marxism-Leninism, setting goals for socialist development, specifying instruments to realize the goals, and designing strategy to achieve these goals (Hsu 1991). These differences are evident in the changing political contexts of the past four decades, which demonstrate essentially a cyclical shift of ideology and power between the radical ideologists led by Mao Zedong and the pragmatic leadership of Liu Shaoqi, Zhou Enlai, and Deng Xiaoping (Skinner 1985; Skinner and Winckler 1969; Keidel 1991). The waxing and waning of these different ideologies, and the coming and going of the decisionmakers who endorsed them, created the different political and economic contexts in which the three development agents identified here found their positions and interacted to shape the spatial economy. To view the complete picture of spatial transformation, we must next consider these forces in their changing historical contexts, and analyze how they combined to alter the contours of the developmental landscape.

3
Maoist Plan-Ideological Space

Major institutional changes initiated in December 1978 at the Third Plenum of the 11th Central Committee of the Chinese Communist Party are now generally considered to represent a significant watershed that divides the history of the People's Republic into two phases. There is, of course, a recognizable continuity that extends throughout the history of socialist development, but overall the kind of ideology and policy introduced since 1978 is considered significantly if not fundamentally different from the kind that prevailed in Maoist China. What then are the differences that exist between the Maoist and the post-Mao eras in terms of development ideology and economic strategy? How has the Chinese spatial economy changed as a result of implementing different strategies?

In analyzing the transition of the spatial economy from the Maoist to the post-Mao era, I have found very useful the conceptual framework developed by Dahrendorf (1968), Johnson (1982), and Henderson and Appelbaum (1992). This framework broadly classifies political economies into four main types:

(1) plan-ideological, which essentially refers to the command economies under classic state socialism
(2) market-ideological, which is epitomized by the American and British economies, particularly under the Reagan and Thatcher 'new right' administrations
(3) plan-rational, which can be applied to the East Asian economies, notably Japan, Singapore, South Korea, and Taiwan, where private ownership and free competition are the norm but the state directs the economy through tax policy and infrastructure development
(4) market-regulatory, which refers to some European economies such as those in Germany and the Netherlands, where the state regulates the

economic parameters but allows private companies to make decisions on investment, production, and distribution (Oi 1995:1134; Dicken 1994:114).

Although some unique characteristics distinguish the Chinese approach to national development from that of classic state socialism, the political and economic system that existed in the Maoist era fits into the 'plan-ideological' quadrant of the above framework. After 1978 economic reforms, China started to move in a new direction to develop a 'socialist market economy with Chinese characteristics.' This gradual process of transition is far from completion, as the influence of Maoist ideology remains surprisingly strong and the single Communist party power structure is still in place. However, in view of the increasingly important role played by free market forces, and the gradual retreat of the planned economy, a new type of mechanism that resembles the market regulatory system appears to be taking shape in post-Mao China.

The Chinese system of socialist development in the Maoist era originally resembled the Stalinist model in the Soviet Union, because that model was the only one available for China to emulate. This system of state socialism was characterized by the undivided power of the Marxist-Leninist party, which is strongly committed to the Communist ideology, and a central command economy dominated by the state or publicly owned sector (Kornai 1992:361). The Chinese system has, however, demonstrated its own special characteristics, which are not necessarily the same as those of the Soviet model. After all, national development in China had been guided by Mao Zedong Thought, which claims to be a doctrine that creatively integrates the general principles of classic Marxism-Leninism with the special practice of the Chinese Revolution. To better understand the dynamics of spatial transformation under Mao, it is necessary to examine the Chinese 'plan-ideological' system and analyze how this system was translated into the creation of a socialist spatial economy.

In dealing with the issue of national development, the Maoist regime that came to power in 1949 appears to have been guided by the fundamental tenets of Marxism and nationalism. More specifically, the new regime emerging from a 'semi-feudal' and 'semi-colonial' nation, and nurtured by some radical Marxist ideology, had two broad goals in mind. The first goal was to create a socialist society where people could live in harmony. The second was to develop an industrialized economy, in which military power could be strengthened and national integrity be secured. The first goal was to be realized by launching numerous political campaigns to promote correct Communist ideology and to transform the Chinese people into ideal 'Communist men,' who are supposed to be unselfish, self-denying, and eager to offer mutual assistance (Eckstein

1977:34). The second goal was to be accomplished by creating a centrally controlled economic system, then believed to be the most rational and efficient way to organize production. These goals gave rise to the two most salient features of national development in the Maoist era.

Communist Ideology

Emphasis on the role of ideological commitment to socialist development is probably one of the most significant features that distinguishes the Maoist regime from the post-Mao leadership. This approach cannot be understood without referring to the perennial debates among top Chinese leaders over whether socialist development should be achieved by promoting correct Communist ideology or by using modern technical expertise. In Chinese, this controversy is usually summarized as the paradox of *hong* (red), which refers to Communist ideology, versus *zhuan* (expert), which means professionalism or technical competence.

Despite some historical fluctuations and an occasional shifting of emphasis, socialist China under the radical Maoist regime had been generally dominated by a focus on correct Communist ideology rather than on technical expertise or professionalism. The idea was essentially derived from the Communist guerrilla warfare experience, which demonstrated that a Red Army of peasant folk equipped with primitive weapons but strongly infused with revolutionary zeal could eventually conquer a nationalist army empowered with superior machine guns, tanks, and a modern airforce. Mao and many of his radical associates believed that the key to military victory is not weapons but a correct ideology that infused the people who operated the weapons. Similarly, the key to successful economic development should not be technical know-how or modern machinery. Instead, the correct Communist ideology could indoctrinate and motivate the masses, who could then be transformed into a tremendous and inexhaustible source of energy sufficient to conquer nature and change the world (Eckstein 1977:46). In the Maoist vision, correct ideas or 'redness' is an absolute prerequisite for socialist development that should be stressed at all times, even at the cost of professionalism and technical competence. This development ideology is vividly described by a Chinese slogan popularly circulated in the countryside during the Mao's years: 'We would rather have our farm occupied by socialist straw than by capitalist seedlings' (*ningyao shehuizhuyi de cao, buyao zibenzhuyi de miao*). Implied in this slogan is the idea that the socialist principle should always be upheld when deciding the type and method of agricultural production, even when the capitalist option looked more attractive than the socialist one.

The emphasis on political correctness rather than modern technology significantly affected the performance of the Chinese national economy, as evidenced in 1958 by the Great Leap Forward program, which was essentially a failed attempt to seek rapid industrialization through propaganda campaigns and mass mobilization. The Maoist ideological approach to national development also found its way to alter the economic landscape. One important tenet of classic Marxism is to restrict trade and commercial activities because, in the Marxist vision, commerce is a process of 'unproductive exploitation' in which a portion of the surplus value yielded by manufacturing is appropriated (Solinger 1985:195). Commercialism is thus logically linked with capitalism and rejected by the Communists. Following this Marxist vision, the Chinese leadership under Mao considered trade and commercial activities to be the 'tail of capitalism' that should be cut. Marxist-Maoist negativism toward commerce means that those geographical areas such as the lower Yangtze and the Pearl River Delta, where the tradition of trade and commercial production was strong, suffered the most because they were unable to perform those economic functions for which they have comparative advantages. It also means that those cities and towns developed as local marketing centres but not for political or manufacturing purposes would have to be phased out or transformed because of their 'undesirable' economic foundations.

A Centrally Planned System

The second salient feature characterizing socialist development in the Maoist era is the establishment of a centralized economic system in which production targets, resources and raw materials, and output sales are all controlled or planned by the central state. The development of this system is closely related to the aforementioned 'correct Communist ideology.' In a manner similar to other Marxist-Leninist parties, the Communist regime under Mao was faithfully committed to two grand revolutionary missions: to abolish private ownership, and to eliminate classes, income inequality, and social disparity. These ambitious missions can only be possible after the allocation of resources and distribution of outputs are in the hands of the state.

Immediately after the Communists took power in 1949, the state confiscated all property owned by the previous feudal lords, nationalists, and foreign capitalists. This confiscation was followed by land reform in the countryside to gain the support of the peasants, and by a campaign to merge public and private ventures in the city as a means to eliminate the private sector. It did not take long for the state to install an urban economy dominated by the public sector, but it took a few more years to develop a similar system in the countryside.

Although land was confiscated from landlords of large holdings and redistributed to poor individual peasants, small-holding private farming remained an economic form contradictory to the Communist ambition of eliminating private ownership. Not long after land reform, massive campaigns of collectivization were launched in the countryside. Farmers were asked to form teams, brigades, and eventually communes so that production activities could be conducted on a collective basis.

The commune system that existed from 1958 to 1984 was characterized by Mao as 'big and public' (*yida ergong*). A typical commune had as many as 3,000 to 5,000 farm households, allowing for mass mobilization to carry out large-scale projects of reclamation, flood control, and irrigation. The system also included many elements of a publicly owned enterprise controlled by the state. Under this system, production decisions were made primarily according to the needs of the central state. In a manner similar to factory workers in the city, farmers were told what to produce, how much to produce, what materials to use, and where and at what prices farm output should be sold. As production was conducted on a collective basis, income or output was redistributed according to the number of 'work points' (*gong fen*), which were calculated primarily on the basis of the time units committed by an individual farmer. This work-point redistribution system was designed to prevent any income inequality that might result from different levels of productivity. However, it gave no incentive to those who were willing and able to do a better job. This system, characteristic of the Maoist ideology of egalitarian socialist development, is usually described as 'eating from the big pot' (*chidaguofan*), which means everyone gets the same benefit regardless of different individual abilities and effort. In any event, the commune system was essentially an extension of the urban planned economy into the countryside. Politically, it allowed the central state to maintain direct and effective control of the vast countryside from the top down to the very bottom. Economically, it ensured grain tax collection to support the state's pursuit of rapid industrialization, modernization, and national defence.

Collectivization of agricultural production was accompanied by reorganization of the marketing system. Beginning in 1957, free markets in both the cities and the countryside were closed by the state. In the countryside, a new state-monopolized system of 'unified procurement and sale' (*tongguo tongxiao*) of farm output was installed. This system was complemented by an urban 'household registration system' (*hukouzhidu*) that rationed for urban residents the supply of foodstuffs, edible oil, cotton, and other daily used goods. By the late 1950s, a centrally planned economic system that controlled the production, circulation, and consumption of goods and services in both the city and the countryside was established.

The centrally planned economic system enabled the socialist state to manipulate the political economy to fulfil its grand Communist revolutionary goals. However, it left little room for local initiative and global forces to play a part in national and regional development. Economic decisions made at the top level according to the needs of the central state were often unsuited to local conditions and insensitive to changes in market demand. Moreover, the command economic system was developed and operated primarily on the basis of vertical administrative linkages, without considering the need for horizontal exchange. This rigid system of decision-making was established to ensure instant and efficient implementation of economic policies made by the central state. One important effect of this system was the creation of a settlement hierarchy dominated by a few large cities and integrated by vertical political and economic linkages.

To some extent, the form of the Chinese command economic system and its spatial consequences bore some interesting resemblances to those of the Fordism that prevailed in the Western world in the early part of this century (Lipietz 1987; Scott 1992). Although developed in a different context and operating at a different level, both the Chinese command economic system and the Fordist regime of mass production are characterized by a welfare state, a rigid structure of economic organization, an inability to respond to constant market change, and an emphasis on continuous capital accumulation. If Fordism was seen as a form of labour organization within a firm, the Chinese centrally planned system can be seen as a means of economic organization within a nation. What is more interesting is that these two systems both resulted in a process of concentrated industrial growth, a settlement hierarchy dominated by large cities, and the emergence of some great industrial regions.

Investment Strategy

The plan-ideological nature of the Maoist political economy was ultimately manifest in the transformation of the economic landscape. Before we move on to unfold its spatial manifestations, one more crucial element needs to be scrutinized. It concerns the Chinese investment strategy that links to plan and ideology on the one hand, and spatial development on the other. In many ways, the investment strategy adopted by the state in the Maoist era resembled the one already implemented in the Soviet Union under Stalin, although some technical modifications were made later in the 1960s when the two Communist regimes split up for ideological reasons. This Stalinist-styled investment strategy is characterized by its bias toward industrial growth at the cost of agricultural development and urban consumption. More specifically, the strategy was designed to seek high rates of industrial growth through heavy investment in some key industries, such as the production of iron, steel, and other capital equipment. The idea was to maximize industrial production capability through the mobilization of all necessary resources to enable China to break out of the vicious circle of backwardness in a short time. This idea was especially evident in the 1950s when Mao called upon the Chinese people to 'overtake the UK and catch up to the US' (*chaoying ganmei*) in industrial production within fifteen years. The failure of the Great Leap Forward campaign in the late 1950s and the break-up of the Sino-Soviet Communist alliance in the early 1960s resulted in some reformulations of this development strategy. The new approach, summarized by Mao as 'self-reliance, strive hard' (*zili gengsheng, jianku fendou*) and 'prepare for war, prepare for famine, and serve the people' (*beizhan, beihuang, weirenmin*), emphasized economic independence and national security. However, the long-term objective of seeking rapid industrial growth by maximizing the production capacity of heavy industry remained unchanged.

The causes and consequences of implementing a growth-oriented and heavy-industry-based investment strategy are complicated. This strategy obviously owes its ideological origins to Marxist thought, which tends to overemphasize production of capital goods at the expense of consumer goods. The strategy also clearly stemmed from Maoist nationalism, which sees rapid industrialization as an essential condition for national defence and war preparation. Needless to say, the implementation of this strategy was facilitated by the centrally planned economic system.

What then are the consequences of implementing this growth-oriented strategy? From the geographical point of view, three distinct spatial relationships have emerged. First, as the focus of development was mainly on resource-oriented heavy industry, the strategy was logically translated into a spatial investment policy in favour of those regions that are rich in mineral and energy resources. Geographically, these regions are primarily located in the northeast, north, and southwest of China (see Map 1.1). With its generous natural endowments of iron ore, petroleum, and coal, plus a well-established economic infrastructure, the northeast became the first focus of state industrial investment in the 1950s. This emphasis was complemented by the development of some major industrial facilities in North China, particularly in the provinces of Hebei, Shanxi, and Inner Mongolia. After military tension between China and the Soviet Union built up in the 1960s, the geographical focus of state investment shifted to the 'Third Front' in the southwest, particularly to the interior provinces of Sichuan and Guizhou, whose locations are better protected from possible foreign military attacks and where a good resource base is available to facilitate the continued development of basic industry (Naughton 1988; Cannon 1990; Fan 1995). By comparison, other Chinese regions that are either poor in the deposit of industrial resources or vulnerable to potential foreign military attack were relegated to the bottom of the state's investment list. Guangdong Province is one such unfortunate region, because its frontier location is viewed as insecure, and because until recently no major industrial natural resources had been discovered there.

Second, under the Stalinist development strategy, the type of industry that received the most attention from the state was capital intensive in nature. Given that the amount of capital available was very limited, and the intention was to seek rapid growth in a short time, the state would be irrational to spread investment all over the country. Instead, a sensible approach would be to concentrate the limited amount of capital in a few large projects that could generate instant and visible economic effects to inspire further development. This approach had been adopted by Chinese economic planners in the 1950s, when state-funded industrial develop-ment primarily took the form of 'key projects' (*zhongdian xiangmu*). These key industrial development projects were small in number but each of them absorbed a sizeable amount of capital. During the First Five Year Plan (1953-7), for instance, the state concentrated 20 billion yuan in the devel-opment of 156 key industrial projects, each of them absorbing capital ranging from 10 million to 600 million yuan (Chinese Academy of Sciences 1987:12).

Moreover, the development of capital-intensive industry usually requires sophisticated technology, which is available only in a few selected places, primarily for historical and geographical reasons. The automobile industry, for instance, was better developed in the city of Changchun in the northeast. The defence industry was relatively advanced in the large city of Chongqing in the southwest. Iron and steel production was traditionally strong in the city of Wuhan in Central China. To efficiently develop technologically sophisticated industry, the state had to concentrate investment in a few large cities where a good technological infrastructure was in place to accommodate modern indus-trial development. During the 1950s, many key industrial projects funded by the state were located in large cities in northeastern, northern, and eastern China (Wu 1967:56-59). The legacy of this concentrated and city-based industrial development remains very noticeable in the present-day economic landscape. The three largest integrated iron and steel manufac-turing complexes, for example, are located in the three large cities of Beijing, Shanghai, and Wuhan.

For a short time in the late 1950s, particularly during the Great Leap Forward, attempts were made to foster the pace of iron and steel produc-tion by setting up backyard steel furnaces all over the country. This prim-itive style of industrial production did not, however, absorb substantial capital from the state. The campaign was launched to mobilize all indige-nous resources as a complement to state-funded and capital-intensive industrialization. The result of the campaign turned out to be disastrous, as the iron and steel produced in rural backyards was almost all useless. This failed attempt became a lesson to show that modern technology and technical expertise cannot be completely compensated for by correct

Communist ideology. It also reinforced the more sensible approach of concentrated industrial growth in large cities.

In any event, the capital and technological intensiveness of heavy industry, which was the emphasis of state investment in the Maoist era, resulted in a development biased toward large cities. State investment favoured large cities, particularly those large cities with important manufacturing facilities. During the late 1960s and early 1970s, when national defence and military preparation were deemed necessary, there were attempts to decentralize industrial facilities from large cities to some inaccessible small settlements. This costly approach has not, however, significantly altered the existing urban primacy.

Finally, implementation of the growth-oriented strategy deepened the divide between city and countryside, in spite of the rhetoric of anti-urbanism and rural development. To finance an industrial program that aims at high-growth rates and focuses on the capital-intensive sector inevitably requires a large amount of capital. In a situation where few external sources were available to provide loans or investment, the only way to generate funds was through high rates of domestic savings. Urban consumption, therefore, had to be kept at a very low level. It also meant that the state would not be able to commit to capital and technological investment in the agricultural sector. In other words, improvement in agricultural production had to be made by increasing labour input rather than by increasing capital investment. To translate this economic logic into spatial terms, the people in the city and those in the countryside had to be separated for managerial purposes. On the one hand, existing cities had to be protected from immigration and their population growth limited so that the cost of providing and maintaining urban facilities could be minimized. On the other hand, a large number of peasants had to be kept in the countryside to produce foodstuffs and sell them to the state at low prices to feed the urban industrial workforce. How then could these ideas by realized?

The artificial rural-urban separation was made possible by two creative management systems unique to Maoist China. First, the whole Chinese population was divided by the central state into two parts: 'agricultural' and 'non-agricultural.' The former usually refers to those people who were engaged in agricultural production in the countryside or in the suburbs. The latter refers to urban residents who had officially approved urban status and who were granted by the state the privilege of obtaining rationed foodstuffs, medical care, educational services, and other urban facilities for free or by paying very low prices. Most of the privileged non-agricultural population were in the industrial labour force. Second, cities and towns were divided into two types: 'designated' and 'undesignated.' No settlement could become a city or designated town without the approval of the State Council or the appropriate provincial authorities. Officially approved cities are known as designated cities (*jianzhi shi*) and approved towns as designated towns (*jianzhi zhen*). Only the designated cities and towns are included in state budgetary expenditures. When a settlement receives an official urban designation, it receives a great deal more financial support from the state than do other places for maintaining urban services and facilities as well as for urban housing development and other types of construction (Ma and Cui 1987; Chan and Xu 1985). The combined outcome of these two innovative managerial systems, created because of the imperative of rapid industrial growth, was a clear spatial separation between urban and rural settlements.

The Unevenness of the Maoist Spatial Economy

Having analyzed the plan-ideological nature of the socialist political economy, and its growth-oriented industrial investment strategy, we can now unfold the complex pattern of the Maoist developmental landscape and see how South China fits into that spatial setting. In general, the spatial economy created by the Maoist regime was characterized by an uneven distribution of population and economic activities. This uneven space evolved as a combined outcome of the forces identified here, despite the rhetoric of egalitarianism. At the macro-level, the unevenness of the Chinese spatial economy is most evident in the spatial distribution of population and economic activities of three geographical regions: the coastal, inland, and border zones (Kirkby 1985:139; Leung 1990:404; Fan 1995:423). However, the disparity among these three zones is actually the result of different geographical and historical conditions that existed long before the Communists took power. Although national economic development under Mao contributed little to altering the existing disparity among the three zones, it would be unfair to blame the Maoist regime for exacerbating these deep-rooted economic and geographical differences. Besides, whether it is necessary and possible to reduce the inequality among these three macro-regions remains debatable (Zhou 1995:127). What the Maoist regime has been directly responsible for is the uneven spatial economy of a relatively developed industrial north and a relatively underdeveloped agricultural south.

The North-South Contrast

Traditionally, South China, particularly the Pearl River Delta region, has relied on the agricultural sector as the backbone of the regional economy. A subtropical climate characterized by warm temperatures and abundant precipitation has made the region especially suitable for the production of paddy rice, pond fish, and tropical fruits. The existence of a navigable river system and a great number of marketing centres across the region has provided excellent conditions necessary for transporting and marketing farm produce. The economic outcome of these favourable natural endowments has been a commercialized agricultural system that was developed as early as the Ming Dynasty (1368-1644 AD).

Unfortunately, the comparative advantages that South China had in commercial agriculture did not fit into the development agenda of the Maoist regime. As discussed, commercialism was considered an exploitative activity from which capitalism might germinate and therefore should be restricted in accordance with the orthodox tenets of Marxism. The agricultural production on which South China had relied also received little attention from the central state in Beijing, which was occupied with the ambitious task of producing as much iron and steel as possible to surpass the UK and catch up to the US in a short time. Moreover, South China is a region that has long suffered from a shortage of energy and mineral resources. There was no industrial resource base to be tapped. It would not be economical to transport energy and raw materials from the north to support industrial growth in the south. Even worse, South China is located on the frontier, facing the capitalist territories of Hong Kong, Macao, and Taiwan. It was seen as extremely dangerous to locate any key industrial project in a region like South China, which was considered vulnerable to possible military invasions by the enemies overseas. For these reasons, South China in general and the Pearl River Delta in particular had been excluded from major state investment for industrial and urban development for most of Mao's years. In this situation where state monopoly precluded any role for local initiative or foreign assistance, South China had few opportunities for economic development. The consequence was a relatively underdeveloped south relying on collective paddy rice cultivation at the subsistence level.

For the north of the country, the story is quite different. The northeast and the north had considerable advantages for the development of a modern industry. The region is distinguished by its abundant deposits of raw industrial materials, a well-established infrastructure, and a relatively skilled manufacturing labour force. Major mineral and energy resources in the northeast, such as iron ore, coal, and petroleum, exceeded other regions in both quantity and quality. More than 60 percent of the total coal deposit in the country was found in North China, 10 percent in the northwest, and only 1.8 percent in the south. Shanxi Province alone contained one-third of China's coal, while Inner Mongolia had 25 percent of the nation's total verified reserves (Pannell and Ma 1983:196). The spatial distribution of iron ore demonstrated a uneven pattern similar to that of coal. Over 50 percent of the verified reserves of iron ore was found in the three provinces of Liaoning, Hebei, and Sichuan. Some of them were located in clusters, which made it possible to build modern integrated iron and steel industrial complexes. The iron and steel complex of Anshan in the northeastern Liaoning Province, for example, was developed by mobilizing iron ore from Benxi and coals from Fushun, both located nearby, to support efficient integrated manufacturing. Similar resource wealth existed in North China, particularly in the provinces of Shanxi, Hebei, and Inner Mongolia.

The region also inherited a relatively advanced electrical and transport infrastructure from the colonial powers (Japan and Russia) that occupied the territory between the two world wars. Most manufacturing and transport facilities were confiscated by the Communists from the foreign capitalists or Chinese warlords, and they became the backbone of state-owned industry. The existence of a large manufacturing labour force was also an important asset for socialist development. In the orthodox Marxist vision, factory workers are ideologically superior to the peasant folk, and they are seen as more determined to make a revolutionary commitment than the latter. As factory workers own nothing except their own labour, they were deemed more ready to sacrifice for the revolution's sake. Farmers, however, were portrayed as those who joined the revolutionary force with hesitation, because they still owned land. The large number of factory workers available in Northeast and North China therefore became a valuable resource that the Communist regime could count on, not only for realizing its ambitious goal of rapid industrialization, but also for winning the battle against private ownership.

Moreover, Northeast and North China were strategically more important at the time than was South China. The northeast area had better connections with the former Soviet Union, which provided the necessary aid to initiate industrial growth. North China was essentially the hinterland of the national capital. These factors all combined to place Northeast and North China at the top, well above South China, of the state agenda for concentrated industrial growth. This fact is evident in any analysis of the geographical distribution of industrial development initiated by the socialist state. Of the 156 key industrial projects funded by the state during the First Five Year Plan (1953-7), 86 projects were located in Northeast and North China, while only 18 were in Central and South China (Chinese Academy of Sciences 1987:12). The result of implementing this concentrated investment strategy has been an uneven economic landscape dominated by the north and the northeast. Despite the shifting investment focus to the 'Third Front' region later in the 1960s, the economic disparity between the north and the south remained unchanged until economic reforms were initiated in the late 1970s.

Large Cities versus Small Towns

The uneven nature of the Maoist plan-ideological space is also manifest in the structuring of cities and towns. As briefly mentioned, the urban system in the Maoist era was characterized by the overarching importance of a few primate cities and the erosion of small towns, particularly rural marketing towns. In interpretations of the Chinese urban development experience, the Chinese model of urban growth is variously attributed to the strategy of 'development from above' or to the practice of 'development from below' (Hansen 1981; Stohr 1981; Wu 1987). Both approaches in fact shaped the pattern of the Maoist urban system.

'Development from above' or the 'top-down' approach has the consequence of strengthening the dominant position of large cities in the urban hierarchy. As discussed, one of the important features of the Maoist political economy is a rigid centrally controlled economic system that operates on the basis of vertical linkages rather than horizontal exchange. This system is only operational when a well-established settlement hierarchy is built so that decisions made at the top level can be passed down efficiently and effectively. As the economic system was monopolized by the central state, an appropriate settlement system should naturally be structured around a few key cities that act as the centres or nodes for transmitting state decisions. The number of key cities should also be limited, because too many key cities would jeopardize the instant transmission of central state decisions.

This strategy was especially important in the 1960s and 1970s, when the central state often had to make and pass down drastic decisions immediately to cope with crises resulting from domestic political turmoil and foreign military threats. The outburst of a coup attempted by Mao's associate in 1971, for instance, resulted in a crucial decision from Zhou Enlai, premier of the State Council, to immediately close all airports in key cities across the country. This decision would not have been implemented effectively and efficiently if the key cities were not state controlled, or if too many key cities existed all over the country. These key cities included the three Special Municipalities (Beijing, Tianjin, and Shanghai), provincial capitals, and some prefectural cities. In addition to their political functions, they were often the focus of industrial development. One of the nation's largest integrated iron and steel industrial complexes, the Capital Steel Co. (*Shoudu gangtie gongsi*), is located in Beijing. Similarly, other large cities, including Shanghai, Tianjin, Nanjing, Wuhan, Chongqing, and Shenyang, have all been the focus of industrial development. A comprehensive study of the economic structure of Chinese cities revealed that manufacturing activity dominated the economies of large Chinese cities (Pannell 1989). The decision to locate major industrial facilities in these cities was apparently in order to use their existing technological and urban infrastructure. Needless to say, such an unbalanced approach to industrial growth in favour of large cities greatly reinforced their dominant role in the urban hierarchy.

Moreover, key Chinese cities were seen as the key nodes in China's transportation network. Until recently, all major airports were located in the national and provincial capitals. In the 1950s and 1960s, railway extension aimed to connect all provincial capital cities with the national capital (Leung 1980). Clearly, the idea was to secure territorial integrity and ensure efficient implementation of political and economic decisions. This approach to transport extension oriented toward key cities also significantly facilitated the growth of large cities.

While the role of large cities was undoubtedly strengthened by political, industrial, and transport developments, a different kind of process was under way at the bottom of the settlement hierarchy. In the countryside, following the socialist collectivization and communization campaigns of the 1950s, a new policy was implemented to stress food grain at the expense of market farming. All over the country, commodity production and sideline business were restricted, because of perceived ties with capitalism. This policy clearly stemmed from Marxist-Maoist anti-commercialism. It also represented a necessary action to ensure that sufficient food grain be supplied for not only the sake of city-based industrialization but also the preparation for war. The implementation of this policy effectively limited the source of rural commodity exchange and undermined the marketing function of many Chinese towns. The direct assault on the economic base of Chinese towns occurred in 1957, when all rural markets were closed and replaced by a state-monopolized system of procurement and marketing. The commercial activities of towns were thus limited to only one or two state-run cooperative stores offering a few basic goods for peasants and townspeople. As commodity exchange and private trade had traditionally been the chief economic function of towns, the socialist attack on commercial activities both in towns and in the countryside inevitably led to a downturn in town development. Moreover, under the city-based and industrial-biased investment policy of the time, virtually no opportunities existed for towns to receive funds from the central state for expanding or maintaining town facilities. Despite the rhetoric of 'strictly controlling the growth of large cities' and 'vigorously developing small towns,' little evidence suggests that the central state made any investment decision in favour of the growth of small towns. On the contrary, small towns suffered from the city-biased development strategy in the Maoist era and they were unable to recover until after the 1978 economic reforms (Skinner 1985; Fei 1986; Tan 1986a; Ma and Lin 1993).

The combined outcome of two different types of development at the top and the bottom of the settlement system has thus been a distinct urban hierarchy dominated by a few large cities. Despite the declared commitment of limiting the growth of large cities, urban primacy for Maoist China had remained relatively high until new flexible economic policies were implemented in the late 1970s to foster the growth of small cities and towns.

The Rural-Urban Divide

The unevenness of the Maoist plan-ideological space was also mirrored in the sharp contrast of development between cities and countryside. For years, the Maoist spatial policy had been known by its egalitarian approach, which aimed to reduce and eventually eliminate the 'three great differences' between industry and agriculture, city and countryside, and mental and manual labour. With few exceptions, this ambitious objective had never been translated into concrete action. No outstanding evidence shows that the three great differences were significantly reduced in the Maoist era. As revealed by a number of scholars, when Chinese planners set out to design and implement spatial policy, their overriding concern was the imperative of rapid industrialization, not the rhetoric of anti-urbanism or pro-ruralism (Kirkby 1985; Chan 1992, 1994). It was the grand ambition of rapid industrialization that fostered a strict divide between industry and agriculture, and between city and countryside.

For most of the Maoist years, the organization of economic activities was guided by the principle summarized as 'agriculture as the base' (*nongye weijichu*) and 'industry as the lead' (*gongye weizhudao*), a strategy in which the purpose of agricultural production was to provide foodstuffs and other materials for industrial growth in the city. This strategy means that a large labour force must be retained in the countryside to strengthen the 'base' of agriculture, particularly the production of food grains. The requirement was effectively enforced by the centrally planned system under which peasants produced whatever the state needed and handed over to the state whatever they produced.

The importance of food-grain production was constantly stressed during the 1960s and early 1970s, when a 'self-reliant' and 'self-sufficient' economic strategy was adopted in the face of a hostile international environment. For strategic and ideological reasons, the emphasis on traditional agriculture left little room for rural industrialization and commercial specialization. This emphasis, along with the restriction on trade, retailing, recreation, and consumption, led the countryside to become persistently rural with no sign of increased industrialization and urbanization. In the early 1970s, the rural communes had managed to use indigenous resources to develop 'five small industries' (*wuxiao gongye*, that is, agricultural machinery, chemical fertilizers, cement, coal mining, and iron and steel). Although such rural industries benefited the commune and provided assistance to farming, they failed to contribute anything significant to rural urbanization.

On the urban side, cities were effectively protected from the influx of rural-urban migration for the purposes of urban manageability and social stability (Kirkby 1985). By restricting rural-urban migration, the state managed to save the cost of providing urban facilities and public services to new urban dwellers. The combined result of a single-sided emphasis on food-grain production in the countryside and protected urban growth was an effective but invisible wall that separated the city from the countryside (Chan 1994).

The state did attempt to reduce rural-urban differences by sending 17.52 million urban educated youth from the cities to the countryside during the late 1960s and early 1970s. However, this transfer campaign was not motivated by a concern for rural-urban disparity. Instead, the notion of 'receiving re-education from the peasants,' like many other rhetorics, was nothing more than a mask. The real motivation was to demobilize Mao's Red Guards, whose violent force was out of Mao's control toward the end of the Cultural Revolution. Evidence revealed by Mao's private doctor lends support to this argument (Li 1994). Moreover, this transfer campaign did not in any way contribute to a significant reduction in urban and rural differences.

Summary

The spatial economy created in the Maoist era, as a geographical manifestation of the plan-ideological nature of the Chinese political economy, included a striking disparity between an industrial north and an agrarian south, between large cities and small towns, and between cities and countryside. In a sense, this spatial unevenness appears ironical, because it stood in opposition to the Maoist declared commitment to egalitarianism and spatial equity. It may also be contradictory to some conventional images that tended to link the Maoist approach with balanced spatial development and the post-Mao approach with uneven polarized growth. However, the result is more understandable after a close scrutiny of the actual forces in operation. Obviously, the unevenness of the Maoist plan-ideological space is the product of a powerful socialist central state that effectively limited the role played by local initiative and global forces. It is also a spatial phenomenon that took place in a particular political, economic, and historical context. As this context changes, the nature and salient features of the spatial economy will also change.

4

Post-Mao Market-Regulatory Space

The Chinese national economy on the eve of the 1978 economic reforms was in crisis. In the words of the Chinese, the economy was 'moving toward the edge of structural collapse' (*zou xiang bengkui de bianyuan*). Endless political campaigns had exhausted the energy of the people. The economy suffered from low productivity, deficiencies, and structural imbalance. In the countryside, the prolonged practice of 'taking grain as the key link for the rural economy' (*yiliang weigang*) limited the variety of economic pursuits for peasants and generated a sizeable army of unemployed estimated at 40 million to 90 million people (Yeh 1984:693). In the city, the unbalanced approach to industrial growth resulted in excess stock of capital goods in the state's warehouses and a severe shortage of consumer goods in the market. Disillusionment with Maoism increasingly built up in the nation and occasionally broke out, as evidenced by the Tiananmen incident in April 1976. Disillusionment alone was unable to force any systematic change, given that the absolute and supreme leadership of Mao was still in place. With the death of Mao in September 1976, a nation tired of turbulent class struggles was ready and prepared to make a drastic change. The arrest of the ultra-leftist 'gang of four' immediately after the death of Mao was widely celebrated all across the country as signifying the end of the prolonged Maoist era of revolutionary ferment and upheaval. However, systematic and structural change of the political economy did not occur until the end of 1978.

Although the transition of the political economy from the Maoist to the Post-Mao era has been a process better described as evolutionary rather than revolutionary, the Third Plenum of the 11th Central Committee of the Chinese Communist Party is generally regarded as the significant landmark leading the nation into a new era. From a historical point of view, the Third Plenum may be no less significant than the Zunyi Conference of January 1935. While the Zunyi Conference led Mao to seize control over the Red Army and the Communist Party, the Third Plenum became a political platform for Deng Xiaoping and his practical associates. It was in the Third Plenum that important ideological and institutional changes brought to a close the Maoist era of revolutionary transformation, and began to lead the country into a new age of post-Mao economic development.

Pragmatism

Perhaps the most distinctive feature of the post-Mao leadership, as compared with the Maoist regime, is its pragmatic attitude toward national development. Whereas Mao saw correct Communist ideology as the prerequisite for socialist development, the post-Mao government under Deng Xiaoping insisted on pursuing alternatives. Deng refused to follow the Maoist doctrine that any economic strategy should be justified by the scripture of Marx and Engels before implementation. He contended that practice is the only criterion to test if the general principle of Marxism is workable. In this vision, the correctness of development practice should not be judged by some conceptual Marxist models. On the contrary, the correctness or suitability of classic Marxism should be judged by real-life practice. The prime concern is thus not the abstract 'correctness' of an economic strategy but the concrete outcomes of implementing the strategy. This pragmatic philosophy is summarized in Deng's well-known saying, 'It doesn't matter if a cat is black or white, so long as it catches mice.' The idea that socialist development should be approached practically, realistically, and flexibly is clearly implied. This pragmatic attitude represents a significant if not a fundamental departure from Maoist ideology. Deng's pragmatic philosophy ultimately resulted in the relaxation of control by the central state over economic affairs, which in turn made room for local initiative and global influence. Pragmatism has not only restructured the political economy but also redrawn the contours of the national economic landscape.

As expected, this shift of emphasis in development ideology would naturally be followed by changes in economic strategy. However, this outcome is not exactly the case in post-reform China. Economic reforms did not occur as a result of careful planning by the central state. Instead, the Chinese experience should be properly described as 'reforming without a blueprint' or 'groping for stones to cross the river' (*caize shiban guohe*) (Naughton 1995:5). This fact is evident in any close look at how reform measures were introduced. In many cases, reform innovations did not come from the top level of the state. They were actually initiated at the grassroots and eventually got approval from the central state. The implementation of the 'household responsibility system' is a perfect example (Ash 1988:534, 554). In many areas, the role played by the central state has been reactive rather than pro-active. The state has not actively introduced new reform programs as the general public might have imagined. Instead, it played a regulatory role, making constant adjustments according to changing circumstances.

To argue that reform is a bottom-up spontaneous process without a pre-developed strategy does not necessarily mean that the central state has not consciously responded to changing circumstances. Nor does it deny that policy changes made by the central state have been essential to the nation-wide promotion of reform initiatives. The changing nature of the function of the central state from interventional to regulatory is by itself a result of conscious self-adjustment. Moreover, a number of political and economic adjustments made by the central state at the macro-level have indeed proven instrumental to the transformation in the political economy.

Specifically, three important adjustments made by the central state under Deng's pragmatic leadership created profound impacts not only on the transition of the national economy but also on the spatial rearrangement of production activities. First, the modernization program was reoriented from the goal of rapid industrial growth to one of realistic economic development. This reorientation was clearly motivated by the pragmatic thinking that upgrading people's standard of living would be more meaningful than blindly seeking some unrealistic high-growth rates for industrial production. Second, the role of central planning was significantly reduced, coupled with the introduction of free market forces in the national economy. This transition of the economy from planning to market coordination is in fact a realistic and practical solution to the problem of structural imbalance between production and consumption inherited from the Maoist era. Finally, the function of the central state was changed from interventional to regulatory as a result of the decentralization of economic decision-making. This functional change was made to stimulate local development and raise individual productivity. Together, these three pragmatically motivated macro-level adjustments have formed a new operational setting under which the national space-economy is profoundly transformed.

From Growth to Development

The first important change resulting from Deng's pragmatism was the creation of a relaxed political environment and realistic goals for national economic development. To restore the national economy after the ten catastrophic years of the Cultural Revolution, the Deng regime initially reset the agenda for the party and the nation. Internal power struggles among different party lines and national revolutionary upheavals lasting for three decades were finally brought to an end. Whereas the previous agenda was dominated by class struggle and political purification, the prime task of the new government was to seek modernization and economic development.

The program of modernization was also fundamentally reoriented and reorganized. Instead of targeting some unrealistic growth rate or inflated iron and steel output, the objective now was to increase productivity, raise per capita income, and upgrade the standard of living.

One of the most important goals for national economic development is to lead the Chinese people to reach 'a moderately high level of living' (*xiao kang shui ping*) by the year 2000. Although the precise meaning of the goal remains unclear and its feasibility questionable, the fact that economic development no longer aims at high rates of industrial growth represents a remarkable departure from the Maoist approach. The industrial program under Deng also targeted for the year 2000 the quadrupling of the 1980 gross industrial and agricultural output value, and an increase in the per capita GNP from US$253 in 1979 to US$1,000. This target was modest in comparison to that for the Great Leap Forward in the 1950s. Moreover, the specification of per capita GNP and income as the goal for national development, which had never occurred before, indicated that the prime concern had shifted from growth rates to real improvement in the standard of living.

This retargeting of the Chinese modernization program happened just a few years after the reorientation of international aid to Third World nations from economic growth to development. It remains unknown whether the Chinese approach was a result of indigenous rethinking or foreign influence. It is certain, however, that the retargeting of the modernization program ultimately led China to move out of the impasse of 'involutionary growth' to 'transformative development' (Huang 1990). More importantly, to adopt a practical, realistic, and meaningful goal for economic development results in abandoning the previous strategy oriented to high-growth rates and biased toward heavy industry. From a geographical point of view, this move has had very important spatial implications as it opens up tremendous opportunities for those regions that were unable to grow simply because they did not fit into the Maoist political agenda.

From Central Planning to Market Coordination

The shift of emphasis from 'brain-cleaning' to nation-building, and from growth to development, has proven to be essential for the transformation of the spatial economy. However, the more direct and immediate impact on the economy created by Deng's pragmatism has been the reorganization of the centrally planned system and the introduction of free market forces. As discussed, the centrally planned economy developed in the Maoist era was a rigid and arbitrary system under which production targets were set and given by the central state according its blueprint of rapid industrialization and national defence, without considering actual market demand.

For ideological and strategic reasons, people's consumer needs were purposely ignored, and the system ran a course that could be best described as 'production for production's sake' (*wei shengchan er shengchan*). The result was a noticeable gap between what had been produced and what people actually needed. While the production of capital goods increased, there was little improvement in people's standard of living.

The practical and realistic reorientation of the modernization program under the pragmatic leadership of Deng meant that the previously unbalanced relationship between production and consumption, and between arbitrary plan and actual market demand, would have to be corrected so that people could enjoy the benefit of economic development. How was this goal going to be accomplished? What was done under Deng's leadership was simultaneous adjustment in two areas: relax the state control over the market and reduce the scope of planning.

One significant move made by the state was to relax its control on pricing. Commodity prices that formerly were set solely by the state are now allowed to fluctuate according to market demand. To avoid the possible chaos that might be caused by a sudden change in prices, the state introduced a 'dual-track pricing system' (*shuang gui zhi*), under which a single commodity could have both a state-set planned price and a free market price (Ash 1988:546; Naughton 1995:8). In many areas of the Pearl River Delta, however, this dual-track pricing system evolved into one dominated by free market prices.

Apparently, this change was made because pricing often acts as an invisible hand on behalf of market demand. When a commodity is in great demand, its price will go up, which in turn will raise the profits of production and eventually increase the supply of the commodity. By allowing prices to change freely, the imbalance between production and consumption, and between supply and demand, could perhaps be corrected. Moreover, since prices are directly connected with profits, free market pricing could provide better incentives for enterprises to increase productivity, upgrade quality, and strengthen competitiveness.

To relax state control over prices would not, however, have any significant economic effect unless enterprises were allowed to respond to market change. This operational relationship seems to have been recognized by Chinese leaders, who since the early 1980s have managed to orient the planned economy toward the market. The state has substantially reduced the scope of the compulsory plan so that the necessary room could be made to accommodate the new element of market-oriented production. The essence of state-planned production remains, however, although it is now in a compressed and stagnated form. In industry, the production of raw materials, such as petroleum, coal, hydroelectric power, iron, and steel, remains under the control of the state. Large state-owned enterprises that are involved in the production of these key industrial materials still have the obligation to fulfil the state mandatory plan. However, these enterprises are allowed to engage in non-plan, market-oriented production after meeting the targets of the mandatory plan. In other words, a state-owned enterprise can develop a dual-track production system, with one track oriented to the state plan and the other one to the market. Enterprises involved in the manufacturing of consumer goods are free to make their own production plans according to market demand, so long as they hand over to the state a certain amount of their profits. For enterprises in the non-state sector, the ban on private ownership has been lifted. Firms that are collectively or privately owned can set up and enter the economy to compete with the state-owned enterprises. This option opens up tremendous opportunities for numerous small-scale, non-state, or non-plan industries to engage in consumer-goods production, a highly profitable venture previously ignored by the state sector for ideological reasons. The consequence of this relaxation of control over production has been the emergence of state and non-state sectors, or plan and market segments, at the macro-level of the national economy and within enterprises owned by the state.

A similar dual-track economic system has been created in the agricultural sector. While the importance of food-grain production continues to be stressed by the state, its dominance in the rural economy has been considerably reduced, giving way to the growth of market farming. Beginning in 1985, the centrally controlled marketing system, under which the state monopolizes procurement and sale of agricultural products, finally was abandoned after over twenty years. A new dual-track system of production and marketing similar to the one for industry has been introduced. Under the new system, farmers still have the obligation of fulfilling the state quota for the production of grain and cotton. However, these key farm products can be sold to the state at two different prices: one contracted with the state and the other determined by the market (Ash 1988:546). Production of farm commodities other than grain and cotton, formerly restricted by state plan, is now unlimited. Farmers are encouraged to engage in a variety of commercialized agricultural activities to meet market demand. An immediate effect of this approach has been a remarkable transition of the rural economy from one that is dominated by food-grain production into one that relies more on market farming, rural industries, and commercial activities. Within the agricultural sector, the emphasis of production has shifted from food grain to commodities such as vegetables, livestock, aquatic products, and sideline business, because such commodities simply are more marketable and profitable than the former. The growth of rural industrial and commercial activities outside the agricultural sector has been even more phenomenal. The sudden explosion of consumer demand among the people who live in the countryside, after being suppressed for thirty years, has created an enormous market that the traditional state industry is unable to satisfy. Such a hugely lucrative opportunity has been seized by peasants who have set up a great number of 'township and village enterprises' (*xiangzhen qiye*) to run factories, stores, hotels, restaurants, and many other commercial facilities. Consequently, the rural economy has become increasingly commercialized and industrialized.

The emergence of a dual-track economy in both the city and the countryside is no doubt a concrete outcome of thinking that maintains a practical, realistic, and flexible approach to national economic development. This economy has turned out to be more desirable and manageable than those resulting from the 'big bang' approach adopted by the former Soviet Union and some Eastern European countries. The creation of this dual-track economy enabled China to undergo a gradual and evolutionary market transition process without experiencing the pain and chaos caused by abrupt change.

While the transition of the Chinese economy from state planning to market coordination is characterized by the coexistence of plan and market segments, the market segment has unquestionably experienced the most dramatic growth, because it often has higher profitability, greater employment capacity, and better linkages to consumers than the plan segment. It is probably because of these advantages that the market segment has become the economic focus of the state, as is evident in any close analysis of the changing tone of the guiding principle for reform. This principle started with a commitment to 'developing a system dominated by the planned economy and complemented by a market economy.' The dominance of the state plan is clearly underscored here. As time passed, the principle was gradually modified to reduce the dominance of the plan segment. It was rephrased to developing 'a system that integrates both plan and market economies,' giving equal weight to the two components. More recently, the principle changed to developing 'a socialist market economy,' with the emphasis now shifted to the market segment.

The tendency of Chinese planners to eventually develop a market economy that 'grows out of the plan' has been recognized by scholars (Naughton 1995). It remains uncertain whether Chinese planners wanted to give up the plan economy and concentrate on the market segment. It is obvious, however, that the traditional plan economy has been running behind the market segment in productivity, efficiency, and economic returns. The momentum of growth for the market segment has clearly been far greater than that for state plan.

The changing nature of the national economy, from domination by a state plan to a dual-track system oriented toward the free market, has restructured the developmental landscape. As the economy shifts emphasis from planning to market coordination, and from capital-goods manufacturing to consumer-goods production, traditional industrial regions supported by the state plan have started to withdraw from the main stage of national development, giving way to other economic regions that are better suited to commodity production. Because of the dual-track nature of the economy, industrial regions that supply energy and major mineral resources to the state continue to play an indispensable role in national development. However, these regions no longer dominate the economic landscape, as many other new regions have emerged to take the lead in consumer-goods production, market farming, trade, and commercial activities. A similar process of restructuring is taking place in the sphere of urban development. Whereas the urban hierarchy was dominated by a few large cities that functioned as the key nodes in the Maoist centrally planned system, the hierarchy is restructuring as numerous small towns start to grow rapidly, given their revival of marketing, trade, manufacturing, and service activities. The process of spatial restructuring that results from the marketization of the economy has also provoked fundamental changes in the grassroots. As the rural economy becomes commercialized, industrialized, and urbanized, the arbitrary division between city and countryside, or urban and rural settlements, has been increasingly blurred, leading to the emergence of a distinct geographical phenomenon characterized by the coexistence of factories and rice fields, industry and agriculture, urban and rural activities. Before we systematically unfold these complex and fascinating processes of spatial restructuring, another crucial factor for change in the political economy requires special analysis because of its instrumental role in the process of spatial transformation. This factor concerns the decentralization of decision-making that results in the function of the central state changing from interventional to regulatory.

From Direct Intervention to Indirect Regulation

One of the most significant characteristics of pragmatic leadership under Deng is its tacit laissez-faire approach to the growth of the national economy. Instead of keeping a tight control of all economic aspects, the state gives local governments and individuals the necessary autonomy and freedom to make economic decisions so long as the interest of the central state is properly honoured. This approach does not comply with the principles of classic state socialism. Nor does it follow the practice of the Maoist regime. How has this approach come into being? What is the raison d'être for the decentralization of decision-making? More importantly, what are the geographical consequences of the changing role played by the central state?

If the shifting emphasis for the national economy from plan to market was motivated by the intention of correcting the structural imbalance between production and consumption, or between supply and demand, the change in the function performed by the state is motivated primarily by the strong desire to overcome the deficiencies of the inherited Maoist bureaucratic system and to raise local productivity.

Of course, to eradicate bureaucracy and quickly raise productivity is not a new goal of development first articulated by Deng's regime. The previous government under Mao was also very concerned about how to raise the productivity of the nation in a short time. To some extent, Mao's desire to quickly lead the nation out of the vicious circle of backwardness was probably even stronger than Deng's. The 1958 campaign of the Great Leap Forward, for instance, essentially represented a desperate and impatient but determined attempt by Mao to pull China out of miserable impoverishment through what Eckstein calls 'a once-for-all' big-push effort, tapping all the energies and latent capacities of the Chinese people (Eckstein 1977:58).

While the two governments under and after Mao share a common concern for raising the productivity of the nation, they are distinguished by the means they choose to achieve that goal. Mao's commitment to raise the productivity of the nation was based on the principle of socialism. To Mao, development must be achieved with the condition that the benefits of increased production be shared equally among the general population. The overriding socialist principle of equity and social justice should never be secondary to any attempt to raise productivity efficiently. Mao's idea was made operational through the creation of a socialist redistribution system under which the making of economic decisions was tightly controlled by the central state that looked after the welfare of the entire nation. Development was also to be achieved by mass mobilization, exhortation, moral appraisal, and symbolic rewards, because such means can stimulate enthusiasms without causing significant unequal distribution of tangible goods.

Deng, however, refused to adopt Mao's ideological approach to increasing productivity. To Deng, it is unrealistic and premature to emphasize equity and social justice in a country that is still at an early stage of socialist development. Instead, it should be acceptable and even desirable for some Chinese people and geographical areas to become rich first. To arouse people's incentive to produce more, Deng insisted on using material rewards as an indispensable instrument for raising productivity. In dealing with the paradox of 'redness' versus 'expertise,' or ideological correctness versus professionalism, Deng clearly stressed expertise and professionalism. Deng also recognized the existence of different productive capabilities among people, and varying comparative advantages among regions. As well, he prefered organization on an individual basis over mass mobilization. By linking material rewards with individual effort, all individuals would be motivated to realize their various personal strengths to the greatest extent so that economic development could be efficiently achieved.

Mao's ideological approach to economic development based on his warfare experience eventually was proven unrealistic and ineffective. The Great Leap Forward turned out to be a great disaster. Mismanagement coupled with natural catastrophe caused 30 million casualties and left many homeless. Mao himself was forced to step back from the front line of decision-making, opening the way for the pragmatic leadership of Liu Shaoqi, Deng Xiaoping, and Zhou Enlai. By using material rewards and other practical measures, Liu and Deng managed to put the national economy back on track after three years of economic readjustment (1961-3). Unfortunately, the relaxed period of economic readjustment and recovery did not last long, as Mao soon launched the massive political campaign of the Cultural Revolution to recover his lost power and reclaim his supreme leadership. Nevertheless, this short period of economic readjustment served as a valuable pilot experiment to convince Deng that pragmatism does stand as a viable development alternative to Mao's ideological approach. Having witnessed the devastating effects of Mao's approach, and personally experiencing the success of pragmatism, Deng Xiaoping, who came to power in 1978 after ten years of exile in remote Jiangxi Province, was well prepared for a nationwide and long-lasting promotion of his idea of seeking economic development by emphasizing efficiency over equity, material rewards over moral exhortation, and individual performance over mass mobilization.

Rural Reforms
Just as Mao's principle of equity and social justice had to be grounded in state socialism, to operationalize Deng's idea of raising productivity efficiently requires major institutional changes to encourage local and individual development enthusiasm. This process of institutional change started from the grassroots. In making these changes, however, the central state did not seem to have a well-considered strategy or a long-term plan. Nevertheless, the new central authorities under Deng's pragmatic leadership appeared to be prepared to make the necessary concessions to accommodate local development initiatives. The state also seemed to understand that if concessions were to be made, the priority must be the peasants in the countryside who represented the majority of the population and who had lived in misery for years. There are, of course, other important political reasons to start economic reforms in the countryside rather than in the city. For a new government striving to establish and consolidate its legitimacy, nothing is more important than satisfying the pressing needs of people for food. Rural reforms also represented a relatively easy and manageable starting point for the Deng regime, not only

because the necessary experience had already been gained from the successful pilot experiment in the early 1960s, but also because reforms in the countryside normally require less capital than in the city. Thus, it is not surprising that the issue of agricultural production in the countryside received the immediate attention of the post-Mao government.

The process of rural economic reforms has been documented extensively and requires no repetition here (Ash 1988; Yeh 1984; Riskin 1987). However, two important points merit special discussion because they have significant implications for the transformation of the spatial economy. First, rural economic reforms essentially represent a process of decentralizing decision-making in which individual farm households are responsible for their own profits and losses in agricultural production. Central to the reforms is abolishing the socialist collective system and adopting some kind of an output-linked contracted 'household production responsibility system.' Under the new system, each farm household is allocated a piece of land and asked to sign a contract with the local state (more precisely, with a representative of the central state in the countryside), which specifies the quota the farm household has to fulfil in return for land allocation. This new arrangement links material rewards directly with individual performance, because the output produced beyond the quota could be disposed of at will by the farm household. It also gives farmers the necessary freedom of decision-making, because they can now choose to engage in other economic activities as long as the contracted quota, which could be met with money or subcontracted to other farmers, is fulfilled. This system not only stimulates the peasants' incentive to produce more but also facilitates the rationalization of the rural economy on the basis of varying local advantages and individual specialties.

The adoption of the production responsibility system, which originated spontaneously from below, eventually revolutionized the political economy of the countryside as it resulted in the abolishment of two thirty-year-old socialist systems: the collectivized system of commune-brigade-team, and the state monopoly of procurement and marketing of agricultural produce. With the dissolution of all major operating socialist systems in the countryside, the central state is no longer responsible for setting production targets, supplying raw materials, and marketing farm products. Instead, it relies on indirect economic measures to control the rural economy. In other words, the function of the state in rural economic development has been gradually changed from interventional to regulatory. Relaxed control by the central state over the rural economy did not in any way weaken the state's ability to realize its economic interests in local development. On the contrary, it enabled the central state to extend its tax collection more effectively than before to the very bottom of the countryside (Shue 1988).

Second, the introduction of this output-linked production responsibility system has triggered a significant process of commercialization and industrialization in the rural economy. As labour productivity increases, a growing number of farmers have found themselves redundant to agricultural production. These surplus rural labourers, who used to be retained on the farm for involutionary grain production, are now free to develop new occupations according to market demand and their own specialty. Because the economic return from non-farm activities is usually higher than that from traditional rice cultivation, many rural labourers have moved out of the agricultural sector to pursue their interests in such areas as trade, transport, processing, construction, and storage. A particularly important outlet that has attracted many surplus rural labourers is rural industry, most of which involves processing local farm products or manufacturing consumer goods. Building on the early development of 'five-small industries,' Chinese peasants have spontaneously initiated an unprecedented revolution of rural industrialization as a means to raise income drastically and generate employment.

The process of rural industrialization and commercialization is crucial to the transformation of the national economy. Without this process of internal redistribution of occupations, Chinese peasants would have to continue to suffer from the cycle of involutionary growth, or flood into cities to seek employment. Rural industrialization and commercialization thus serve as an avenue leading peasants to turn away from the impasse of involution, and to move into the new era of transformative development.

Agricultural restructuring is, of course, made possible by the relaxation of control by the central state over the rural economy. It has some significant geographical implications. In a manner similar to industrial restructuring, rural economy after reforms has demonstrated a special pattern of growth in which non-farm activities have expanded much faster than the traditional agricultural sector. Geographically, this process of economic restructuring has translated into a spatial process in which growth has concentrated in those places with good conditions for the development of non-farm activities. Such conditions could include natural endowments, historical tradition, a skilled labour force, and proximity to large markets. The lower Yangtze and the Pearl River Delta are two good examples. Moreover, commercialization and industrialization of the rural economy has provided the impetus to greatly facilitate the growth of small towns in the countryside and foster the pace of rural urbanization.

Urban Reforms

A similar process of decentralized decision-making has been under way in the city. Compared with their rural counterparts, Chinese cities may not have experienced very dramatic and successful economic reform, partly because no quick fix can remedy the many structural problems inherited from the Maoist plan-ideological system, and partly because urban reforms usually require some intricate economic measures that are relatively new to the Deng government. Nevertheless, several important institutional changes have proved conducive not only to the transformation of the urban economy but also to the spatial restructuring of production activities.

Essentially, institutional change in the city represents an attempt of the post-Mao government to promote efficient economic development by linking material rewards directly with local and individual performance. The centre of institutional change is the introduction of a 'fiscal responsibility system' (*caizheng baogan*), an urban translation of the production responsibility system already practised in the countryside. Under this fiscal arrangement, an urban enterprise or municipal government is asked to sign a tax agreement with the central state. The enterprise or local government is free to make its own production decisions so long as the tax quota is met. Any remaining surplus can be retained by the enterprise or local government for reinvestment, redistribution, or consumption. Such an arrangement gives enterprises concrete economic incentives to produce goods that are profitable or marketable, because the more profit they make, the more surplus revenue they can retain. The arrangement also makes enterprises and local governments responsible for their profits and losses (*zifu yingkui*), and compels them to overcome the problems of deficiencies and mismanagement inherited from the Maoist system of state socialism. With this system in place, the central state is no longer directly involved in local economic affairs. One significant geographical consequence of this functional change is that cities and regions with better revenue-generating capabilities will grow much faster than those that have long relied on the protection and support of the central state. As suggested by Nee (1989:667), those regions that used to be 'contributors' in the socialist redistribution system before the reforms can now experience net gains over those that were beneficiaries.

Although the central state no longer monopolizes all local economic affairs, it has maintained its function of regulating or coordinating the national economy. One remarkable change made by the central state is its use of banking as an instrument to regulate the growth of the national economy. The central state is not responsible for making detailed decisions about production and distribution, but it remains the sole owner of banks and the largest owner of capital in the country. By adjusting interest rates and bank loans given to local governments, the central state can effectively influence the direction of the national economy. Moreover, major mineral resources and transportation facilities remain in the hands of the state. Tax collection and personnel appointment are two other powerful means the state can rely on to coordinate local economic development. Control over the growth and movement of population is an additional regulatory function performed by the central state that has significant implications for spatial transformation. The 'one-child' policy has been strictly enforced in large cities but implemented in a more lenient manner in the countryside, primarily because rural population growth usually costs the state less in welfare provision than its urban counterpart. Similarly, migration into large cities has remained under the strict control of the state. Illegal migrants to the large cities are denied access to medical care, schooling, and other public utilities. They also face constant threats of deportation and eventually are removed by the city authorities. However, since 1984 the ban over migration from rural areas to small cities and towns has been lifted. Peasants are now allowed to relocate from the countryside to small cities and nearby towns on the condition that they can provide themselves with food grain, edible oil, and other necessities without causing an extra welfare burden on the state. This partial relaxation of control over rural-urban migration has undoubtedly facilitated the growth of small cities and towns and contributed to the restructuring of the settlement hierarchy.

Open Door Policy
The decentralization of decision-making is thus not a process of the central state completely giving up its power. It is rather a process through which the central state makes a trade-off with local people both in the city and in the countryside. For the purpose of arousing local enthusiasms and promoting efficient economic development, the central state can afford to give up some of its power in certain economic areas which are not indispensable and which the central state has never been able to manage well in the first place. This tacit laissez-faire approach toward some selected unimportant economic areas has been translated into a spatial development strategy that Naughton (1995:11) calls 'disarticulation,' a strategy under which several peripheral and unimportant areas have been allowed to go on their own. The first two peripheral provinces that have been freed to go on their own are Guangdong and Fujian, both located in the southern corner of the country far away from the national capital and both poor in major industrial resources. These two provinces were identified in 1979 as the locale for practising an 'open door policy' (*kaifang zhengce*) and developing Special Economic Zones (*jingji tequ*).

The reasons for choosing Guangdong and Fujian as the target region for implementing this special policy are twofold. First, these two provinces are the home of the majority of overseas Chinese and thus have the extensive overseas connections necessary for attracting foreign capital and promoting export production. They are also located in geographical proximity to Hong Kong and Taiwan, which are seen as not only the main sources from which external capital is to be mobilized but also the window through which information about the world market can be obtained. Second, these southern provinces are strategically unimportant to the growth of the national economy. The central state could never afford to lose the modern industrial bases in the northeast, north, and east, but it could allow the two peripheral provinces in the southern corner to undergo some economic experiments. The core of the national economy would remain undisturbed should the southern experiment fail. Alternatively, if the two chosen provinces succeed, they could serve as a model for other regions. This idea may be derived from the ancient Chinese strategy of 'finding direction in the dark by throwing a stone to get a reaction' (*toushi wenlu*). The two southern provinces are thus used by the central state as a 'testing stone' thrown in the dark to get a reaction for determining what should be done next.

The implementation of the open door policy in general, and the development of Special Economic Zones in particular, has been widely documented (Ho and Huenemann 1984; Lardy 1987; Yee 1992; Yeung and Hu 1992). Although the practice of 'disarticulation' is not unproblematic, viewed from the perspective of national economic development, the experiments in South China clearly have not failed. Instead, South China has been able to use the special autonomy granted by the central state to generate phenomenal, self-sustained economic growth and transformative development. The region has become the most attractive destination for foreign and domestic capital investment and for migrant workers from all over the country seeking employment. In a manner similar to the dual-track economy in which growth is concentrated in the free market track, South China has enjoyed practising capitalism in a socialist territory and has demonstrated far greater growth than other parts of the country, including those that served as the core region of the socialist economy.

The Triangular Power Structure
The introduction of market forces and the adjustment of the function performed by the central state have created a new political and economic framework for national and regional development. As the central state removes its constraints on those economic sectors and geographical areas viewed as not indispensable, local initiatives and global capitalist forces have moved in to occupy the vacuum left behind by the retreat of the state. This process is especially evident in the countryside where reduced intervention from the state has enabled local initiatives to chart a new course of rural industrialization and commercialized farming. It is also clearly displayed by the development experience of South China, particularly by the Pearl River Delta region, where local governments have been able to use the special autonomy to reorganize economic activities, set up

a transportation infrastructure, and solicit capital from overseas sources. The state's disarticulation approach to growth in some peripheral regions, such as the two provinces of Guangdong and Fujian, has also opened an avenue for the influence of global capitalism, which primarily takes the form of capital investment, technological transmission, and managerial skills, as well as the spread of Western ideas, behaviours, and lifestyles. The result of restructuring the political economy on both the national and the regional scale has been the emergence of a new triangular operating system in which the central state, local governments, and global forces interact to create a new post-Mao market-regulatory space.

The operational mechanism of the triangular nexus of interaction is extremely complicated. As the function of the central state changes from direct intervention to indirect regulation, the traditional power structure of Maoist state socialism under which 'lower-level governments obey higher-level governments' (*xiaji fucong shangji*) and 'local governments obey the central government' (*difang fucong zhongyang*) has been gradually replaced by a new system in which central state and local governments form a partnership to maximize their own economic interests. This new central-local relationship is a mixture of conflict and collaboration. It involves bargaining, negotiation, and mutual accommodation. It can also result in increasingly more complicated tensions, beyond those that can be easily resolved by the replacement of local cadres. On the whole, however, the new central-local relationship is collaborative and mutually accommodating, because collaboration is in the best interests of both sides.

This new central-local relationship has some significant implications for the transformation of the spatial economy. For example, it has enabled local governments to play a greater, more active role in the development of those regions that are economically, geographically, or psychologically distant from the central state. South China is a prime example of such a region. It has also resulted in the disintegration of the previous rigid political hierarchy, which relied on vertical linkages and nested in a urban system dominated by a few large cities. Horizontal linkages have become increasingly important as local governments are granted greater decision-making power. Such horizontal linkages have to be made on the basis of a settlement system in which cities of different sizes are more evenly distributed than they were before. Finally, the new central-local relationship has been one of the important factors responsible for the recent development of 'feudal lord economies' (*zhuhou jingji*) in some regions where economic barriers were set up to protect local interests.

On the other hand, global capitalist forces have usually made inroads after being successfully adapted to local situations. In bringing global forces into the socialist territory, the central state has been very cautious. Foreign firms are discouraged, if not restricted, from setting up wholly owned ventures in some key economic sectors such as mining, energy generation, and transportation. The operation of joint ventures between overseas and local entrepreneurs will also be subject to close monitoring and regular inspections. For this and other reasons, anyone attempting to run a business successfully must first develop reliable personal connections with local cadres to secure the economic interests of foreign entrepreneurs. Thus, global capitalism has not collided with the local society. Instead, it has been sheltered in some local situations and comes in the form of 'local capitalisms' (Smart 1995). This distinct local-global interaction means that the intrusion of global forces must necessarily be a geographically selective phenomenon. It can only succeed in those areas where local conditions are favourable or accommodating to the practice of global capitalism. Such local conditions could include relaxed state control, geographical accessibility, a good telecommunication and transportation infrastructure, and most importantly the existence of social and cultural ties between foreign investors and the target regions. In the Pearl River Delta, such favourable conditions exist.

Interestingly, the restructuring of the Chinese political economy described here shows some striking similarities, at least on the surface, to changes in Western Europe and North America since the early 1970s, although the contexts in which such changes have occurred are quite different. Well documented by economic geographers, what has taken place in North American and Western Europe has been a fundamental change in the basic industrial organization from the rigid Fordist regime of mass production to a post-Fordist regime of flexible accumulation (Scott 1988, 1992; Storper and Scott 1989). Such organizational changes have been made in response to increasing competition from Japan and the newly industrializing economies arising since the 1970s. It represents an attempt to substantially reduce operational costs and ultimately increase competitiveness, which is similar to the Chinese attempt to eradicate deficiencies inherited from the rigid centrally planned system and to efficiently raise productivity. The transformation of industrial organization from Fordism to post-Fordism has been made possible primarily by the deregulation of a Keynesian welfare state and the development of new technologies that compress time and space, both of which have significantly enabled the flexibility of industrial production. This process also bears a resemblance to the Chinese case, where economic transformation has been facilitated by the relaxation of central state control and the development of the local-level transportation infrastructure. More interestingly, the changing central-local relationship in the Chinese case since the early 1970s is similar to its American and European counterparts, where the central state has turned away from redistribution goals and paid more attention to industrial rationalization (Scott 1992: 221).

What makes the Chinese case distinct from the Western situation is perhaps the dual-track nature of the economy, in which plan and market mechanisms, or state and non-state sectors, coexist, with growth concentrated in the market or non-state sector. As discussed, the rigid Chinese plan-ideological system resulted in an uneven space, dominated by a few large cities and some core industrial regions in the north. For the American and European cases, the spatial consequence of practising the rigid Fordist regime of mass production had been the creation of 'a bipartite structure' characterized by thriving industrial-urban regions (for example, the manufacturing belt of the United States) and a subservient set of agricultural regions (Scott 1992:220). The industrial core 'contained innumerable large urban agglomerations rising out of the industrial base and housing the masses of workers employed in the local areas' (Scott 1988:173). Eventually, the turn to a post-Fordist regime of flexible accumulation gave rise to the emergence of 'new industrial space' that is usually situated away from the former heartland of Fordist production identified by economic geographers (Scott 1988; Storper and Scott 1989). It would be interesting to assess if a similar spatial process has taken place in China as a result of the transition from Mao's radical regime to Deng's pragmatic leadership.

A New Spatial Economy

In unfolding the complex pattern of spatial transformation in post-Mao China, we are confronted with two contradictory assessments already produced by geographers. The first and more popular expectation has been that regional inequality of manufacturing output distribution and personal income has increased since the reforms because the new economic policies obviously opted for efficient polarized growth over egalitarianism and equalized development. The other interpretation suggests just the opposite: the post-Mao economic landscape has become more equalized than the one developed in the Maoist era, despite the changing commitment of the state from egalitarian to selective and concentrated growth (Fan 1995; Lo 1990; Wei and Ma 1996). Judged from a technical perspective, these two opposing interpretations may have been the result of different research methodologies, including different scales of measurement, time periods of study, and variables chosen for assessments.

A close evaluation and comparison of the two visions reveals that the fundamental cause of the difference between these visions lies in recognizing South China as a newly developing region whose growth has effectively reshaped the national economic landscape. Indeed, the key to solving the mystery of post-Mao spatial development is the dramatic economic surge of South China, surpassing the northern manufacturing heartland, and thus significantly reducing the spatial unevenness inherited from the

Maoist radical regime. The spatial issue in China has been not so much the persistent disparity between the east and the west but the shifting focus of development between the north and the south. This new process of spatial restructuring has been recognized by a growing number of scholars (Lyons and Nee 1994; Chen 1994; Fan 1995; Lary 1996; Yeung and Chu 1994; Asia Research Centre 1992). This distinctive pattern of post-Mao spatial development, however, is not an outcome of strategic planning by the central state. Instead, it is an unexpected consequence or a by-product of the state's disarticulation approach to the issue of local economic development.

The New North-South Relation

Among all major regions in the country, South China has probably benefited most from the ascendance of the post-Mao regime, not because the new administration has given it any preferential capital allocation, but because the removal of constraints by the central state has allowed the region to materialize its developmental potentials. As compared with other regions, South China has a number of advantages in developing a market economy. First, it has a well-established tradition of carrying out market-oriented economic activities, including trading, commercialized agriculture, and consumer-goods production. The tradition of trade and marketing originated in the city of Guangzhou (Canton), which serves as the chief economic centre of the region. With its gateway location and favourable conditions for seaport transportation, Guangzhou was the first city in the south to grow as a national centre of international trade, dominating the import and export industries of the entire nation, even before Shanghai and other port cities were developed on the eastern coast (Sit 1985; Lin 1986:9). The existence of Guangzhou as China's southern gateway for trade and transportation had in fact been a major factor explaining why early British colonists chose Hong Kong in the south instead of Shanghai in the east as an entrepôt for shipping opium into China in the mid-nineteenth century. Over the years, the tradition of trade and marketing established by Guangzhou has been carried on and further expanded, giving rise to a distinct commercial identity for the region and its people. This commercial identity is discernible in the common Chinese perception that people from the south are usually very good at business but not as capable as northerners in manufacturing. It is also evident in the choice of Guangzhou since the early 1970s as the site of the biannual national export commodities fair, throughout a period when the Communist ideology of anti-commercialism remained highly lauded.

The functional strength of the south in trading and marketing has been complemented and reinforced by the development of commercialized agriculture in the countryside and the growth of consumer-goods industry in the city. With a natural endowment of fertile soil, warm temperatures, and abundant precipitation, South China has long played a leading role in the production of agricultural commodities such as pond fish, sugar cane, silkworm cocoons, vegetables, and tropical fruits. An abundant supply of agricultural commodities paved the way for the rapid development of modern industry, a process that can be traced back to the turn of the century when factories were set up for manufacturing activities, including canning fish, weaving silk, refining sugar, processing food, making bricks, manufacturing textiles and garments, and processing many daily used goods. Such industries might not have been as advanced as their northern counterparts in manufacturing large and durable machinery, but they were closely linked with local farm commodity production and geared toward direct consumption in the market. South China's historical heritage in trade and commodity production was unable to bring its people substantial profits or economic prosperity during the Maoist years. With an economy now in favour of market operation, the 'seeds of commercialism' that have been deeply sown in the soil of the south can freely and rapidly germinate, bloom, and bear economic fruits.

The second potential economic advantage that South China has over other regions is its geographical proximity to the two major newly industrializing economies of Hong Kong and Taiwan. Historically, Hong Kong and Taiwan were both part of South China. Although the two exclaves were artificially separated from the mainland at different times for political reasons, they have shared a common culture with the mainland, particularly with the two neighbouring provinces of Guangdong and Fujian. The geographical closeness to the two capitalist exclaves used to be detrimental to the growth of South China, because it excluded the region from receiving major capital investment from a central state that was deeply concerned about national security. However, the potential role played by Hong Kong and Taiwan as both the source of overseas capital and the window on world market changes means that a tremendous economic opportunity exists for South China following the renewal of its traditional linkages with Hong Kong and Taiwan. With the state's adoption of the disarticulation approach to economic development in the region, South China has quickly seized the opportunity and turned it into an inexhaustible source of generating wealth and prosperity.

Finally, South China has been blessed by its extensive connections with overseas Chinese who live in North America, Europe, and Southeast Asia. About 80 percent of overseas Chinese originated in Guangdong and the majority of the remainder come from Fujian. That South China has been the main source of Chinese emigration overseas may have been the result of earlier contact with the Western world. It is also closely associated with the existence of nearby Hong Kong and Taiwan, both of which served as stepping stones for emigrants from the mainland to other parts of the world (Skeldon 1994). For whatever reasons, the existence of extensive connections overseas has proven extremely beneficial for economic development in the region. By frequently sending home remittance to support relatives, overseas Chinese have often provided the necessary capital to launch local development projects. This overseas channel of capital mobilization could be crucial to triggering local industrialization and infrastructure development, especially at the initial stage when funds from other sources are not readily available. With a good knowledge of world market demands and local production capacity, overseas Chinese have bridged the local economy and the outside world by directly setting up ventures or by indirectly providing business opportunities to the local people. The frequent personal contacts between overseas Chinese and local people have also effectively filtered into the region not only advanced technological innovations and capitalist managerial skills but also Western ideas, behaviours, lifestyles, and consumerism. However, the existence of overseas connections has not always worked in favour of the region and its people. During the Maoist years, local people with overseas connections suffered severely from all sorts of state discriminations, including the denial of their eligibility to join the army, to take on important posts in government agencies, and to enter some high-ranking universities. Now that the balance of values has shifted from ideological correctness to economic utility, to have overseas connections has become an enviable privilege that can enable people in the south to realize their dream of modernization much more easily and faster than those in other parts of the country.

Eventually, the comparative advantages identified here, plus a location distant from Beijing, have led South China to be chosen as a testing ground for practising decentralized decision-making and developing an open-market economy. This choice has brought unprecedented economic growth and structural change to the region. The process of economic transformation under way in the south began with the establishment of four Special Economic Zones along the southern border near Hong Kong, Macao, and Taiwan as a means to attract overseas capital and promote export processing. This step was followed by the designation of some port cities on the coast as Open Cities and the identification of two central areas as Open Economic Regions, including the Pearl River Delta in Guangdong and the Minnan Delta in Fujian. In all cases, foreign investors were offered investment incentives such as tax concessions, and duty-free import of raw materials and machinery. Such arrangements are not the genuine innovation of the Chinese as they had already been practised in the export-processing zones in Singapore, South Korea, Taiwan, and elsewhere. However, their implementation in South China took place at a time when the Asian Newly Industrializing Economies (NIEs) were desperately seeking ways to reduce labour costs in order to drastically increase their global competitiveness. The opening of South China thus provided a great opportunity to the Asian NIEs for tapping the enormous pool of cheap labour on the mainland.

Many manufacturers from Hong Kong and Taiwan have moved into South China to set up joint ventures and processing firms. From 1979 to 1994, the region received 42 percent of all realized foreign capital investment (US$99.893 billion) that flowed into China (China, SSB 1996:110). Currently, South China contributes over 47 percent of the total export output generated by the whole nation (China, SSB 1995:551). The dramatic growth of foreign investment and export production has created substantial employment opportunities, expanded production capacity, and drastically increased income for the local people, thus significantly raising the national economic rank of Guangdong and Fujian provinces. Whereas in 1978 Guangdong was ranked seventh in gross value of industrial output, it now occupies the second most important position next only to Jiangsu Province (China, SSB 1996:76). Similarly, the rank of Fujian in national manufacturing has risen from twenty-first in 1981 to fifteenth in 1995 (China, SSB 1982:19; 1996:76). Without doubt, South China, particularly Guangdong Province, has successfully moved ahead of the nation in the struggle for economic development and modernization. This region, increasingly integrated with Hong Kong and to a lesser extent with Taiwan, has emerged as one of the most dynamic economic regions in the Asian Pacific Rim, with dynamic growth comparable to Japan, Singapore, South Korea, and Taiwan during their early rapid economic development (Vogel 1989; Kwok and So 1995; Sung, Liu, Wong, and Lau 1995).

South China is, of course, not the only economic region that has benefited from the ascendance of the post-Mao regime. Other Chinese regions with comparable advantages in developing an open and free market economy have also experienced rapid growth. These regions include the lower Yangtze Delta, Shandong Peninsula, and some metropolitan regions on the eastern coast. Geographically, these regions are located either along the coast with good transportation facilities to allow export processing, or near some large cities that provide great demand for market-oriented economic activities. Historically, many of these regions had deep roots in carrying out trading, market farming, and consumer-goods production. The transition of the economy from plan to market coordination in these regions is easier than in other parts of the country. Wherever it takes place, the development of regional economies in the post-Mao landscape has been selective. It is characterized by a shifting focus from planned to market operation, and from rigidity to flexibility. It involves deregulation by the central state to make room for local initiative. It is driven by the desire to raise income, productivity, and profits. However, the growth magnitude of these regions has not been as great as that of the south, probably because they do not have South China's autonomy and overseas connections to facilitate economic exchange with the outside world.

While regions in the south and on the eastern coast have vigorously upgraded their positions in the national economy, the growth of the manufacturing heartland in the northeast and the north appears sluggish if not stagnant. In the Maoist era, the industrial regions in the north and the northeast were able to dominate the Chinese production space, primarily because their resource condition and existing manufacturing capacity fit into the state's agenda of rapid industrialization. These regions enjoyed support from the state and were protected by the planning system from free competition. With the mechanism of the economy now shifted from central planning to market coordination, the privileges accorded to the north and the northeast have gradually eroded, leading the regions to confront tough challenges they have never experienced before.

As the principle of manufacturing changes from 'production for production's sake' into 'production for profit,' regions in the northeast and the north face a very difficult task of industrial reorientation. Many industries there produce raw materials or basic machinery unsuitable for direct consumption in the free market. To reorganize production according to market demand will require a fundamental conversion of the existing manufacturing facilities and retraining of the labour force, which are extremely difficult if not impossible goals. Yet without direct linkages to consumption, profitability would be lower, material rewards would be fewer, production incentives would decrease, and productivity would drop. This dilemma is compounded by the gradual withdrawal of direct support from the state, which is now limiting its commitment to the central plan and encouraging the growth of a market economy outside of the plan. Moreover, these industrial regions have been exposed to strong competition from other parts of the country, including numerous township and village industries and newly developing industry in the south. All of these new rivals are capable of pursuing market-oriented production, because most of them are small-scale, self-managed, and easily adapted to rapid changes in the market.

After three decades of manufacturing guided by the central plan, it is simply too difficult for people in the northeast and the north to 'swim in the billowing ocean of the commodity economy.' Because of these constraints, the traditional industrial heartland in the northeast and the north is gradually losing its dominant position in the national economy. This is not to suggest that the production capability of the region has declined or output value has dropped. Both the northeast and the north have actually been able to maintain their leading position in the production of capital goods, including energy and mineral materials as well as modern heavy

machinery. However, they have been left behind by the south and the eastern coast in terms of the expansion of the regional economy, particularly
in the growth of outputs, income, and employment. As the concentration
of growth shifts from the plan to the market segment, the geographical
centre of growth has moved from the traditional industrial heartland of the
northeast and the north to the new production space emerging in the
south and on the eastern coast. Consequently, the spatial unevenness created by the Maoist radical regime has been substantially reduced, especially
at the interprovincial level (Fan 1995; Wei and Ma 1996).

The Restructured Settlement System

The process of spatial restructuring is also manifest in the reorganization
of the settlement system. This reorganization is evident in an examination
of the changes at two levels of the settlement hierarchy. At the upper level,
the system of cities has undergone some fundamental restructuring. At the
lower level, numerous small towns have grown rapidly to take the lead in
the process of rural urbanization. As the state decentralizes decision-
making to local governments, and brings in free market forces to regulate
the economy, the previous rigid urban hierarchy organized by vertical linkages and political functionality has disintegrated and been replaced by a
new system of cities shaped primarily by horizontal connections and economic exchange. One of the most noticeable consequences of this structural change is the rapid increase in the number of cities since the reforms,
partly because rapid economic growth has created a great demand for
urban functions, and partly because the central state no longer has to limit
the number of large cities to pass down urgent political decisions efficiently. It has been recorded that the number of cities in China increased
from 193 in 1978 to 467 in 1990, among which the number of large cities
with over one million people rose from only thirteen to thirty-one (Zhou
1995:291; Xu, Ouyang, and Zhou 1995:493). An outcome of the increasing
number of cities has been a reduced degree of dominance by the capital city
in urban systems, not only at the national scale but also on the regional
level, as has been documented by Chinese urban geographers (Zhou
1995:290; Yeh and Xu 1996:244; Xu, Ouyang, and Zhou 1995:498).

The functions performed by cities have also changed. In the previous
Maoist era, cities were developed primarily as the centre of manufacturing
production and capital accumulation with a minimal degree of consumption (Lo 1987; Pannell 1989; Chang 1981). Under the pragmatic leadership of Deng Xiaoping, the urban economy is no longer tightly controlled
by the central state. The objective of urban development is not simply to
increase savings and industrial output but to generate more income,
profit, and employment. Many cities have moved to explore a variety of
economic opportunities outside the traditional industrial sector. This

exploration has given rise to a diversified and commercialized urban economy that covers a wide range of activities including not just manufacturing but also catering, retailing, banking, tourism, entertainment, and real
estate speculation. The nature of cities has thus changed from simply the
base of manufacturing production to the centre of commercial exchange,
consumption, and recreation. Geographically, this process of functional
change is most noticeable in the growth of cities in the south and along
the eastern coast.

A more remarkable development has occurred on the lower level of the
settlement hierarchy. It involves the spectacular growth of many small
towns all over the country, challenging the dominant positions held by
large cities. The rapid surge of towns in China is generally a recent phenomenon. Before the reforms, small towns never experienced any significant economic development. On the contrary, they suffered severely from
discrimination by the socialist state, which not only gave investment priority to large cities for the purpose of rapid industrial growth but also constantly launched anti-commercialism campaigns to erode the economic
foundation of small towns. The declared official policy of promoting
small-town development was nothing more than a rhetorical mask to
cover the state's real ambition of city-based industrialization. The takeover by Deng's regime brought a new atmosphere favourable to the
growth of towns. Reduced state control over the rural economy allowed
market farming, sideline business, and non-agricultural pursuits to flourish in the countryside. This development provided tremendous opportunities for agroprocessing, market exchange, and other services in towns.
Eventually, it resulted in the 1981 reopening of free markets in Chinese
towns, after their closure by the socialist state for almost three decades.
Since then, revitalized small towns have again sparked the engine of economic growth and driven forward full speed with the seemingly inexhaustible energy of enterprising townspeople and rural peasants engaged
in industrial, commercial, and other activities (Skinner 1985; Fei 1986; Tan
1986a; Lin 1993; Lin and Ma 1994). Small towns now house a large number of the township and village industries whose output value has reached
almost one-third of the nation's total gross industrial product. They play
a crucial role in the absorption of a great number of surplus rural labourers who have been released from agricultural production as a consequence
of increased agricultural productivity. They have also served as the centre
of modernization in the countryside, exposing the peasants to new ideas,
technology, and the culture of an urban society.

The phenomenal growth of small towns has presented a viable option
of urbanization distinct from but complementary to the norm of city-
based urbanization. This new form of urbanization, called 'urbanization
from below' by Ma and Lin (1993), is essentially a spontaneous, self-

driven, and self-sustained phenomenon made possible by the state's tacit laissez-faire attitude toward the growth of towns. The growing magnitude of this new dimension of 'urbanization from below' suggests that the Chinese urban landscape dominated by large cities is now undergoing profound structural change. The Maoist urban system centred on a few large cities appears to face serious challenges from some powerful indigenous forces emanating from the countryside, giving way to a more balanced settlement structure where cities and towns play complementary if not competing roles in national urbanization and modernization. If the current trend of town growth and the state's liberal policy continue for an extended time, the town-based dimension of 'urbanization from below' will likely expand and, together with the city-based dimension of 'urbanization from above,' will form a distinctive Chinese pattern of 'dual-track urbanization' (Ma and Lin 1993; Ma and Fan 1994).

Blurring the Rural-Urban Division
The process of spatial restructuring extended beyond the reorganization of the conventional settlement system and created a new settlement pattern in the country. As discussed, one of the most salient features of spatial development in the Maoist era was a strict artificial separation of urban and rural residents resulting from the strategic decision to minimize the cost of providing urban facilities for rural-urban migrants. Despite the official rhetoric of reducing rural-urban differences, the rural-urban divide had been deepened by the policies of investment, pricing, and migration, all designed to exclude peasants from the benefits of urbanization. It was not until the late 1970s that economic concessions and institutional changes were made to give peasants the opportunity of directly taking part in industrialization and urbanization.

As the winds of political decentralization and marketization blew across the country, Chinese peasants started to demonstrate great enthusiasm and individual creativity in restructuring the rural economy to raise their income and profits. The process of economic restructuring is characterized by a disproportionate increase of more profitable, non-staple sideline and non-agricultural activities, and a simultaneous drop in the proportion of traditional farming. The reduced input of labour and land in farming is offset by increased productivity. Such a reorganization of economic activities has been carried out through a rational division of labour. For a typical farm household, the contracted farmland is usually looked after by the housewife or the elderly, whose task is to ensure fulfilment of the grain quota for the state, while the husband, son, and daughter engage in more lucrative non-agricultural endeavours. A similar division of labour also occurs on a seasonal basis. Often peasants will concentrate on agricultural production during the planting and harvest seasons when the demand for labour on the farm is high. They will then shift to non-agricultural pursuits for the rest of the year, thus raising income and maximizing profits. The result of this new division of labour, motivated by profit-seeking, has been a mix of agricultural and non-agricultural occupations in one family.

The growing mix of economic pursuits is even more apparent where the rural economy has developed beyond traditional staple production to incorporate a variety of industrial, commercial, transportation, and service activities. This trend is especially obvious in the rapid development of rural industry that is mostly small scale, labour intensive, and locally driven. The locations of rural industry are invariably in the countryside near the farms and villages. This industry has provided the most effective means for peasants to acquire employment outside the agricultural sector but within their living sphere. Its rapid growth in the countryside has enabled peasants to 'leave the soil but not the village' (*litu bulixiang*) and 'enter the factory but not the city' (*jinchang bujincheng*) (Ho 1994; Byrd and Lin 1990). It has also created a new settlement form in which industrial and agricultural or urban and rural activities take place side by side. This new form does not fit the classic description of 'urban' or 'rural' settlement, but it displays characteristics of both types. This special pattern of settlement transformation has been called 'rural-urban integration' (*chengxiang yitihua*) by scholars and local people in China. It is most noticeable in those areas extending between or around large metropolitan centres where rural-urban interaction has been most intensified. In general, the pattern of rural-urban integration appears to confirm the conceptual model of an 'extended metropolitan region' (*desakota*) in many Asian countries (Ginsburg, Koppel, and McGee 1991; McGee and Robinson 1995).

The driving forces behind the new settlement form of rural-urban integration are complex. They include state deregulation of the rural economy, the introduction of a free market mechanism, agricultural restructuring, an inflow of transnational capital, and the development of a regional transport infrastructure. Whatever the cause, the ongoing process of rural-urban integration has fundamentally changed the unequal relationship between industry and agriculture, and between city and countryside. It has significantly blurred the rural-urban divide that characterized the Maoist developmental landscape. The 'invisible wall' built up by the Maoist regime to separate city and countryside for three decades has been torn down spontaneously by Chinese peasants.

Summary

The Chinese spatial economy under the post-Mao pragmatic leadership has demonstrated some striking features that are significantly different from those associated with national development under Mao. Whereas in the Maoist era the spatial distribution of economic activities was concentrated in the northern industrial heartland and in a few large cities, the post-Mao developmental landscape is characterized by the remarkable upsurge of new production space in the south, the phenomenal growth of small towns, and the rapid urbanization of the countryside. Consequently, spatial inequality, which the Maoist regime failed to overcome, has been substantially reduced at the interprovincial level.

The movement of the national economy toward balance or equality has not been the result of any active government programs or the direct intervention of the central state. On the contrary, it is made possible primarily by reduced state control over the local economy, which has enabled local initiative and global market forces to play an active role in national and regional development. It is this new operating mechanism in which the function of the central state changes from active command to passive guidance that explains why post-Mao spatial development tends toward balance, despite a state policy seemingly in favour of spatial polarization. It is also the state's disarticulation approach rather than its active intervention that explains much of the locally driven development in South China, the small towns, and the vast countryside.

To some extent, the transformation of the Chinese spatial economy since the reforms resembles the process of post-Fordist flexible production under way in Europe and North America. Both cases are shaped by state deregulation and growing marketization. They are characterized by the emergence of new production space outside the manufacturing heartland that engages in small-scale and flexible specialization. However, the Chinese case is distinct from its American and European counterparts in at least two respects. First, the Chinese spatial economy is distinguished by its dual-track nature, particularly the coexistence of state and non-state sectors, or plan and market segments. Second, the operating forces in the Chinese context appear to have emanated primarily from the countryside, whereas the American process of flexible production is driven more by the relocation of firms from the metropolitan cores to some suburban areas. It does not seem reasonable to suggest that the Chinese and Americans have influenced each other in reorganizing their spatial economies to respond to global interdependent development after the Cold War. It seems fair to argue, however, that the logic of spatial transformation, particularly the operating mechanism of local-global interaction, has some regularity and commonality that deserve further investigation.

Over the past four decades, the leaders of socialist China were preoccupied by the challenging task of rapidly developing a modernized economy while maintaining social stability and national integrity. The two developmental goals of modernization and social stability have been shared by both the Maoist and the post-Mao regimes, although such goals were defined as high-growth rates of industrial output by the former, and specified as increased income and employment for the general population by the latter. While the two governments have common goals to pursue, the actual measures they have taken are significantly different.

The Maoist regime put its faith in the seemingly unlimited power of development that might be generated by ideological commitment. It relied on an imposed centrally planned economic system that was supposed to be capable of using human and natural resources most rationally and efficiently. Such a plan-ideological system brought with it a rigid and centralized power structure typical of state socialism, where the socialist state monopolized all political and economic affairs, and left little room for local initiative, individual creativity, and global market forces to play any active role in national and regional development. The economic consequence of practising state socialism had been a pattern of 'growth without development,' characterized by considerable industrial output growth, with no substantial improvement in productivity and per capita income. The national economy suffered from a severe imbalance between an arbitrary plan and the actual market demand, and between production and consumption. This peculiar system, in which state intervention extinguished local enthusiasm, central planning overtook market regulation, and production overrode consumption, resulted in a process of spatial transformation that created an uneven developmental landscape. Industrial production was highly concentrated in the northeast and the north, where major energy and mineral resources were found. The settlement system was dominated by a few large cities that functioned as centres of industrial production and key nodes for transmitting centrally made political decisions. Cities and countryside were arbitrarily separated to ensure a sufficient supply of food grain and to reduce costs to support urban expansion.

With a keen recognition of the dysfunctionality and deficiencies of the Maoist system of state socialism, the post-Mao government adopted a pragmatic approach to national and regional development. This approach gives professionalism and technological expertise greater weight than abstract ideology or political correctness. It accepts the fact that material rewards are essential to stimulating local enthusiasms and individual creativity. It allows for the decentralization of decision-making and introduces a market mechanism to balance the relation between supply and demand, and between production and consumption. This approach has also created a flexible environment that accommodates the operation of local and global forces. The result has been a process of transformative development featuring the disproportionate growth of market-oriented production and a remarkable increase in productivity, personal income, and general employment. The state's disarticulation approach to some peripheral areas has given rise to the rapid surge of South China, small towns, and the countryside on the economic landscape. Consequently, the uneven structure of the Maoist plan-ideological space has been profoundly altered to make way for the emergence of a relatively balanced, post-Mao market-regulatory space.

The process of economic and spatial transformation under way in China since the reforms is truly fascinating, not only because the enormous population size and geographical scale involved have few parallels in the world, but also because the magnitude of change demonstrated has no comparison in the history of the nation. For an extended time, the Chinese resisted the invasion of global capitalism because they considered themselves as living in the centre of the universe, or 'the Middle Kingdom' (*Zhongguo*), where they could obtain all necessities without having to exchange with the outside world. The door of China was eventually forced open by Western colonial powers in the mid-nineteenth century, when this self-centred and self-isolated nation was brought into the orbit of global capital accumulation. However, because of their strong desire to monopolize the Chinese market and to ship home as many natural resources as possible, Western colonists set up business barriers to block local Chinese firms from entering the sphere of free market competition, and effectively oppressed the growth of indigenous capitalism. Not long after, the Communist regime that took over the country in 1949 ushered in a prolonged period of socialist development in which capitalism, originated indigenously or externally, was unable to take root in Chinese soil. It was not until after the 1978 reforms that, for the first time in Chinese history, local initiative and global forces were able to interact cooperatively, promoting the genuine growth of incipient capitalism. Such local-global interaction has been most intensified in South China, particularly in the Pearl River Delta region, whose geographical proximity to and extensive personal connections with Hong Kong have greatly facilitated the penetration of global capitalism into the socialist territory. The fascinating process by which a southern Chinese regional economy under socialism is transformed by local and global capitalist forces is discussed in the next chapter.

Part 2:
Development of the Pearl River Delta

5
Economic and Spatial Transformation

In my travels around China's southern provinces of Guangdong and Fujian in April of this year, I discovered that roads are being built so fast, in so many new directions, that no maps are accurate. The guidebooks cannot keep up with the hotels and restaurants that have opened – every one is out of date. So are telephone directories and company listings. These explosive changes make China terra incognita ... To make way for cities erected in a matter of months, mountains are being moved, rice paddies filled in, forests cleared ... The dynamo of capitalism has been loosed, and the 'creative destruction' that economist Joseph Schumpeter called the defining feature of nineteenth century American capitalism is on display in the China of 1993. It is a sight the likes of which few people alive today have seen.
– Paul Theroux, 'Going to See the Dragon,' *Harper's Folio*

The Pearl River Delta region is one of the most populous and productive regions in China. Located on the southern coast of the mainland, the delta has long been China's southern gateway for foreign trade and sea transportation, hence, one of the earliest regions open to the outside world. It is also the richest rice bowl in South China. Its subtropical monsoon climate, fertile alluvial soils, and a water system suitable for year-round irrigation and transportation have made it a major agricultural region, leading the nation in the production of sugar cane, pond fish, silkworm cocoons, and tropical fruits such as lychees and bananas. Its traditional leading role in foreign trade and agricultural production has been greatly enhanced since 1978, when liberalized economic policies were implemented. Taking full advantage of its proximity to and excellent connections with Hong Kong, the delta region has been allowed to move ahead of the nation in attracting foreign investment and developing a market economy. Two of China's four Special Economic Zones were established in the region. In 1985, the entire delta was officially designated an Open Economic Region. Growth and development have been phenomenal ever since. Capital investment has been flowing in from Hong Kong and overseas; joint ventures and cooperative trade enterprises have been established and expanded rapidly; numerous bridges, freeways, ports, and harbours are being built; and new farming systems and technology are being practised. The delta, increasingly intertwined with Hong Kong, has quickly emerged as one of the fastest growing and most dynamic regions in the West Pacific Rim.

Areal Extent

The growth and development taking place in the Pearl Delta region has received much scholarly attention. A considerable number of publications have been generated in Chinese and English to document the significance of the economic and spatial transformation of the delta region (Sung, Liu, Wong, and Lau 1995; Johnson 1992; Vogel 1989; Sit 1984; Lo 1989; Chu 1996). Despite the outpouring of scholarly work on the subject, demarcation of the Pearl River Delta region remains vague and confusing. As described by a geographer from Hong Kong, 'different authors, different government departments and even different time periods tend to have different definitions of the boundary of the [delta] region' (Li 1989:38). Indeed, from available literature, it seems that there are as many definitions of the Pearl River Delta as articles and monographs on the subject (Chu 1996:470; Li 1989:38; Lo 1989:295; Yeh, Lam, Li, and Wong 1989:1; Wong and Tong 1984:3-5; Johnson 1992:185; Leung 1993:277). The delimitation of the Pearl River Delta is no less obscure and confused among local Chinese geographers. It has actually been a subject of a long-lasting and unresolved scholarly debate among Chinese geographers since the 1930s (Chen 1934; Wu and Zeng 1947; Huang and Zhong 1958; Zhong and Li 1960; Xu 1973; Xu, Liu, and Zeng 1988; Miao, Shen, and Huang 1988). Despite the great variety of conceptions of the Pearl River Delta, three main definitions can be identified in the existing literature.

The first definition was proposed and popularized by local Chinese geographers based on physical geography considerations. The underlying principle of demarcation is the natural mechanism of delta formation, essentially a result of the interaction between tide and stream flow. It was suggested that the boundary of the delta should go as far as the places where the stream water meets and interacts with the ocean tide. Such places of tide-river interaction usually take the form of what has been called a transitional zone (Wong and Tong 1984:10). It has been generally accepted that the upper reach of this tide-river interaction extends to Lubao of Sanshui xian (county) to the north, Shilong zhen (town) of Dongguan shi (municipality) on the east, Linyoungshia of Zhaoqing shi in the west, and Tan Jiang (Tan River) of Kaiping xian in the southwest (Miao, Shen, and Huang 1988:1-3; Xu, Liu, and Zeng 1988:31; Wong and Tong 1984:5). The Pearl River Delta thus delimited covers an area of 17,200 square kilometres. Map 5.1 outlines the area of the Pearl River Delta according to the physical delimitation.

This conception of the delta, while geographically sound, has gained little popularity outside the academic community, because it is inconsistent with the administrative boundaries of the participating municipalities and counties. Many counties on the periphery of the delta are divided by the boundary imposed by geographers, which leads to problems in data gathering and research.

A second definition of the Pearl River Delta region was officially established in January 1985. For the purpose of deciding which counties could offer preferential treatment to foreign investors, the Zhujiang (Pearl River) Delta Open Economic Region was officially demarcated. This region, covering 22,800 square kilometres, included four municipalities (Foshan, Jiangmen, and two former counties, Zhongshan and Dongguan) and thirteen counties (Doumen, Baoan, Zengcheng, Panyu, Nanhai, Shunde, Gaoming, Heshan, Xinhui, Taishan, Kaiping, Enping, and Sanshui, which was initially excluded but added in 1986). This official definition of the delta is now commonly called the 'Inner Delta' or 'Smaller Delta' (*xiao sanjiaozhou*) by local officials and researchers as well as by Western scholars (Xu, Liu, and Zeng 1988:32; Chu 1996:472; Vogel 1989:161; Lo 1989:298; Pan, Cao, and Yu 1991:145).

Map 5.1

Pearl River Delta as defined according to its physical geography

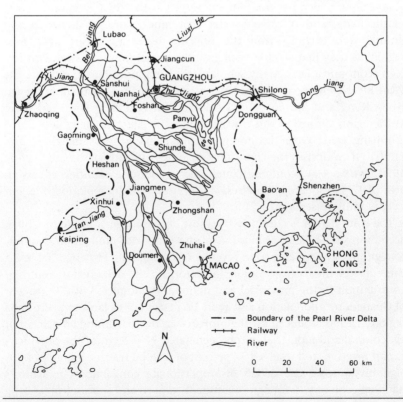

Source: Chu 1996:471; Wong and Tong 1984:5.

A third definition of the Pearl River Delta is virtually an expanded version of the second one. In an effort to hasten the economic growth of the mountainous area surrounding the delta region, the State Council announced in November 1987 that the previously designated Zhujiang Delta Open Economic Region would be expanded by three municipalities (Qingyuan, Huizhou, and Zhaoqing) and eight counties (Huaxian, Chonghua, Huiyang, Huidong, Boluo, Gaoyao, Sihui, and Guangning), which were mostly in the relatively underdeveloped mountainous area. The Inner Delta or Smaller Delta was thus expanded into a 'Greater Delta' (*dasanjiaozhou*), covering 45,005 square kilometres, and containing seven municipalities and twenty-one counties (see Map 1.2). This latest official definition of the delta region has now been employed by local governments, statistical bureaus, and academics as well as the public media.

It should be noted, however, that the official designation of the delta region, either the 1985 Inner Delta or the 1987 demarcation of the Greater Delta, was mainly for the purpose of attracting foreign investment. No effective regional authorities have been set up to govern development within the demarcated delta region. More importantly, Guangzhou, which has traditionally served as the chief economic centre of the delta, and the two Special Economic Zones of Shenzhen and Zhuhai, were excluded from the officially demarcated Pearl River Delta region, as they had already been granted special power to deal with foreign economic affairs. Many scholars felt it inappropriate to separate the Pearl River Delta region from Guangzhou, Shenzhen, Zhuhai, and even Hong Kong and Macao (Chu 1996:476; Zheng 1991). The Pearl River Delta region defined in this study includes the latest officially designated 'Zhujiang Delta Open Economic Region,' Guangzhou, Shenzhen, and Zhuhai. It comprises the eight municipalities officially designated before 1990, two Special Economic Zones, and twenty-one counties; covers an area of 47,430 square kilometres; and houses a total population of about twenty million (see Map 1.2).

The delta region is currently the most developed area in South China, accounting for a large proportion of the industrial production, exports, foreign investment, and retail revenue of Guangdong Province. Table 5.1 lists some of the basic economic indicators for the delta region and their contribution to the provincial and national economies. With 33 percent of Guangdong's population and about 26 percent of its land area, the delta region in 1990 produced for the province an overwhelming 68 percent of agricultural and industrial output, generated 77 percent of the province's export revenue, and received 77 percent of its total realized amount of foreign investment. The delta's disproportionate contribution to China's national economy in foreign investment, export, and subtropical farming products such as sugar cane, pond fish, and fruit is also remarkable (see Table 5.1). When the selected economic indicators are calculated on a comparable per capita basis, the delta clearly stands as one of the most advanced economic regions in the country, far ahead of provincial and national averages on almost all indicators (see Table 5.2).

It is acknowledged that the 'Zhujiang Open Economic Region,' Guangzhou, Shenzhen, Zhuhai, Hong Kong, and Macao have increasingly integrated to form a social and economic entity. A meaningful analysis of the development issues of the delta region would not be possible without taking into account all of the participating urban centres and their interaction. For the purpose of data comparability, however, the Pearl River Delta region defined in this study will exclude Hong Kong and Macao, because of the existing differences in the economy and the system of data gathering. The exclusion of Hong Kong and Macao from the statistical analysis of this book is by no means an oversight of their importance in the development process of the delta region. The social and economic impact of Hong Kong and, to a lesser extent, Macao, on the mainland side of the delta region, has been tremendous. This impact will be fully addressed in this book. The emphasis will be, however, on assessing the impact of Hong Kong on the recent development of the

Table 5.1

Basic economic indicators for the Pearl River Delta, 1990

Items	Unit	Delta[1]	Delta as % of Guangdong[2]	Delta as % of China[3]
Total population[a]	million	20.75	33.21	1.85
Area	thousand km^2	47.43	26.66	0.49
GVIAO[b]	billion yuan	113.82	68.88	6.51
Rice	million tonnes	6.72	35.43	1.50
Sugar cane	million tonnes	8.26	39.44	16.44
Fruit	million tonnes	1.38	42.10	9.04
Fish	million tonnes	0.84	40.57	7.44
Retailing	billion yuan	45.62	62.30	6.51
Export	billion dollars	8.18	77.44	17.34
Realized foreign investment	billion dollars	1.57	77.41	18.95

[a] Total population includes migrants who have resided in the region for ten months or longer. For the definition of total population, see Guangdong, Statistical Bureau 1991a:503.
[b] GVIAO stands for Gross Value of Industrial and Agricultural Output. Data are in 1980 constant prices.

Sources:
[1] Guangdong, Statistical Bureau 1992b:65-7.
[2] Guangdong, Statistical Bureau 1992b:71-2.
[3] China, State Statistical Bureau 1991a:2.

delta, not on tracing the development processes of Hong Kong and Macao, which have already been well documented.

Geographical Context and Historical Background

As discussed in previous chapters, the transformation of a regional economy is the complex outcome of the interaction of various geographical, historical, and social forces. The Pearl River Delta region is no exception. First of all, geographical factors are fundamentally important, as they form the physical setting within which growth and development take place. The Pearl River Delta is geomorphologically formed by sediment carried down by the West, East and North rivers on their way to the South China Sea. The delta is distinctive in that it is not a single piece of extensive flat plain of low relief. Instead, it is a composite of several basins drained by a number of rivers and their tributaries. The confluence of the rivers and the conjunction of their basins take place near Guangzhou (Canton), the biggest urban centre of the region. This composite delta is interlaced with many rivers and their branches, most of which are navigable year-round. Such a special physical setting has made waterways the major traditional means of transportation. It also has necessitated the construction of numerous bridges as door-to-door highway transportation has become prevalent. The issue of transport development will be discussed in greater detail in Chapter 7.

The delta is also distinguished by its natural endowment, which is most favourable for agricultural production. With a subtropical location, the

Table 5.2

Selected economic indicators for the Pearl River Delta in comparison with Guangdong Province and China, 1990

Items	Unit	Delta[1]	Guangdong[2]	China[3]
Population density	persons/km^2	437	351	119
Non-agricultural population				
% of total	percent	36.80	23.65	19.42
Per capita GVIAO[a]	yuan/person	5,488.09	2,645.17	2,756.25
Per capita income[b]	yuan/person	3,425.56	1,812.60	1,250.73
Per capita export output	dollars/person	394.20	169.20	54.30
Per capita realized				
foreign investment	dollars/person	75.49	32.39	8.99

[a] Data are in 1980 constant prices.
[b] Data are in 1990 current prices.

Sources:
[1] Guangdong, Statistical Bureau 1992b:65-7.
[2] Guangdong, Statistical Bureau 1991a:40-7.
[3] China, State Statistical Bureau 1991a:2, 18, 82, 102.

delta enjoys warm temperatures (21-22 degrees Celsius yearly average), abundant precipitation (1,600-1,700 millimetres annually), and ample sunshine (solar radiation 110 kilocalories per centimetre annually), the result being a year-long growing season suitable for double-cropping rice. Such favourable conditions, combined with superior fertile soil and a well-established waterway system for transporting and marketing farm products, have led to the development of an intensive farming system, capable of supporting a dense and rapidly growing rural population.

While the delta is rich in agricultural resources, it has for many years suffered from the absence of major mineral deposits such as coal, iron ore, and other raw industrial materials. It was not until fairly recently that petroleum potential was discovered underneath the nearby South China Sea. The delta's shortage of industrial mineral resources, a factor usually overlooked by many previous studies, has suggested that the region is essentially agriculture oriented, that large-scale urban manufacturing is difficult to create, and that industrialization must be primarily a small-scale and rural-based phenomenon. This special combination of natural endowments in the delta region has a striking similarity to the combination described by Fei Xiaotung in his rural industrialization study of Wujiang xian in southern Jiangsu Province (1986). The importance of agricultural development and rural industrialization in the delta region will be further discussed in Chapter 6.

The unique geographical location of the delta also merits special attention. By virtue of its location, the delta is relatively remote, situated at the southern end of the mainland far away from the political centre of Beijing or the economic centre of Shanghai. This remoteness is further underscored by the existence of a wide range of high mountains (*nanlingshan*), which physically separates the delta from the vast territory of the nation. The physical barrier of the mountains may not be a major concern in the face of modern telecommunications and transportation, but it has been a crucial geographical factor underlying the historical development of the region for thousands of years (Lary 1996).

The important implications of the remoteness of the Pearl River Delta should never be underestimated. It is such remoteness that has given the local people considerable flexibility in seeking development, and, in some circumstances, the possibilities of rebellion or revolution, as evidenced by the 1910 republican revolution led by a delta native, Dr. Sun Yat-sen. On the other hand, this frontier location has made the delta serve as China's traditional southern gateway for foreign trade and sea transportation. Historical records indicate that Guangzhou, the biggest port city in the delta, was one of China's earliest trade outlets, with a history that can be traced back to the Qin Dynasty (221-206 BC) (Lin 1986:7). More importantly, the delta's coastal frontier position and its proximity to Hong Kong and Macao have enabled it to become the first region to benefit substantially from the 1979 introduction of China's open door policy. In the words of the local people, the delta is China's 'window to the south wind' (*nanfeng chuang*), bringing the fresh air of capitalism into a country under socialist rule. The unique geographical location of the delta region, relatively remote from China's heartland but close to Hong Kong, suggests that the economic, social, and cultural impact of Hong Kong on the delta's development is no less significant, if not greater, than the impact of Beijing. In fact, since 1979 'the wind from the south' has become even stronger than 'the wind from the north.' This issue will be further addressed in Chapter 8.

The physical environment as outlined here has provided the local people of the delta with opportunities which have, throughout history, been fully explored. An intensive farming system existed long before the turn of this century, based on the production of rice, sugar cane, mulberry, and fruit, as well as silk cocoons and pond fish (Zheng 1991:42). The increased production of farm commodities gave rise to thriving industries such as sugar refining, fruit processing, fish canning, silk and textile manufacturing, paper-making, and ceramic/porcelain manufacturing, which were primarily based on local resources. Thus, despite the fact that the delta region was not richly endowed with energy and mineral resources, agricultural and aquatic production provided a raw material base diversified enough for flourishing small-scale manufacturing. The increased output in farming and manufacturing had in turn led to increased trade and prosperity as marketing and transport networks developed and grew. The concentration of commercial activities and the specialization of production facilitated the agglomeration of population in towns and cities. By the late nineteenth century, the delta region had become one of the most urbanized economic regions in China, next only to the lower Yangtze region (Skinner 1977:211-49).

The 1949 victory of the Communist revolution led the delta region to enter a peculiar stage of socialist development. The trade embargo imposed by the United Nations in the early 1950s deprived the delta of its role as the nation's leading outlet for trade and export. In the countryside, traditional commercialized agricultural production, which had been the economic base of the delta region for decades, was terribly disrupted by the socialist campaigns of collectivization and communization. With a single-sided rural economy overly focused on paddy rice cultivation, farmers had few commodities to sell in markets or process in local factories. The regional economy was further assaulted in the late 1950s when the state implemented the policy of the 'unified procurement and distribution' (*tonggou tongxiao*) of grains, cooking oil, cotton, and other essential materials. The policy made a state monopoly for trade, which further eroded the naturally evolving commercial system. During the Great Proletariat Cultural Revolution (1966-76), rural sideline production and commercial activities were seen as 'the tails of capitalism' that must be and were indeed cut. Many rural factories, workshops, and retail establishments were forced to shut their doors as they lost the necessary resources and market for production.

In the urban areas, investment in manufacturing and infrastructure was limited, partly because of the vulnerability of the delta's frontier position to perceived naval attack, and partly because of Mao's declared commitment to eradicate the rural-urban and coastal-interior differences. Consequently, the average annual rate of urban growth of the Inner Delta region from 1957 to 1978 was a meagre 0.75 percent, even lower than the national average (Xu and Li 1990:53). Free rural-urban migration was next to impossible because of the strict household registration and grain-rationing system. Border control was tight and those who attempted to escape to Hong Kong or Macao were prosecuted and given life sentences. This scenario was typical of regional development in the delta before 1978. As Vogel comments, 'That the Inner Delta, long known for its commercial dynamism, could be transformed so completely into the same system of collective communes, state commerce, and tight border controls as that put into effect elsewhere testifies to the strength of the Communist control system at its height' (Vogel 1989:164).

The death of Mao in 1976 and the subsequent demise of the ultra-leftist radical leadership in the late 1970s opened a new chapter of genuine development for the Pearl River Delta. The opening of Guangdong and Fujian provinces for foreign investment, the establishment of four Special Economic Zones (two of which are in the delta), and the recent designation of the Zhujiang Delta Open Economic Region all provided great impetus for the delta, not only to resume its traditional position as China's largest trade outlet, but also to play a new leading role in attracting capital investment from Hong Kong and overseas. Meanwhile, a series of new economic policies were implemented, which have given the local people greater flexibility to produce according to market demand and provided incentives to those who are willing to work harder for more material production. The original collectivized agricultural production system was abandoned and replaced by a new responsibility system under which farmers fulfil output quotas contracted with local authorities and can then keep or sell all products above the quotas. The commercialization and diversification of agriculture were encouraged and local markets were revitalized. Sizeable capital was mobilized through various public and private channels and directed to the construction of housing and transportation infrastructure. Consequently, the regional economy of the delta has experienced unprecedented growth and restructuring.

Economic Growth and Structural Changes

That the Pearl River Delta region is one of China's fastest growing economic regions has become a fact that is widely recognized and documented (Sung, Liu, Wong, and Lau 1995; Vogel 1989; Johnson 1992; Lo 1989). Documentation of the delta's unprecedented economic growth, however, has not been systematic, consistent, and convincing, partly because the definition of the delta region varies considerably among different researchers, and partly because the data are in different measurements from various sources, and therefore not easily compared. The Chinese authorities have published a number of statistics showing the fascinating expansion of the delta's regional economy in GNP, GDP, and national income. Such data have to be used with care, however, as the Chinese statistics of GNP (*guomin shengchan zhongze*), GDP (*guonei shengchan zhongze*), and national income (*guoming shouru*) are not exactly the same statistics as those used in the West, albeit the terms are seemingly identical. Moreover, the production figures in these key economic indices are frequently counted in current prices and in Chinese yuan, which are distorted by inflation and hence not comparable over time.

A more appropriate index for measuring economic growth is the Gross Value of Industrial and Agricultural Output (GVIAO or *gongnongye zhongchangze*). Table 5.3 lists the statistics for a number of key economic indices of the Pearl River Delta region and compares them with those for Guangdong Province and for China. It is clear from Table 5.3 that the Pearl River Delta has experienced spectacular economic growth since 1980. Its industrial and agricultural output has expanded at the extraordinary

Table 5.3

Annual growth of output value for the Pearl River Delta in comparison with Guangdong Province and China, 1980-90

Items	Delta[1] %	Guangdong[2] %	China[3] %
GVIAO[a]	19.24	17.1	11.2
Industrial output value[a]	21.23	19.7	12.6
Agricultural output value[a]	6.86	7.6	6.4
Export output value[b]	29.28	17.0	13.1
Realized foreign investment[b]	31.54	25.2	22.8
Retailing exchange[a]	20.08	18.3	7.3

[a] Raw data are in Chinese yuan at 1980 constant prices.
[b] Raw data are in American dollars.

Sources:
[1] Guangdong, Statistical Bureau 1992b:65-7.
[2] Guangdong, Statistical Bureau 1992a:48-9.
[3] China, State Statistical Bureau 1991c:21-3.

annual growth rate of 19 percent, which is much faster than both the provincial and the national averages. The most dramatic growth occurred in the inflow of the realized foreign capital investment and export of local products, recorded at an astonishing annual increase of 31 percent and 29 percent respectively. At the local level, the rapid revitalization of commercial activities, trading, and market exchange has also been remarkable. Data on total retail value show an annual increase of 20 percent over the past decade, which almost triples the national average. There is little doubt that the delta region has moved ahead of the whole nation in attracting foreign investment, practising with free market forces, and achieving a remarkable start-up for its regional economy.

The dramatic growth of the regional economy has brought significant improvements in productivity, employment, and income, as indicated by the data listed in Table 5.4. Again, the most impressive increase occurred in the realized amount of foreign capital investment and export production on a per capita basis, but the improvements in rural per capita income and per capita industrial and agricultural production are also significant. The employment rate, defined as the proportion of the total population between the ages of sixteen and sixty-five who were employed, shows a slight increase, probably because of the continued expansion of the aging population, teenagers, and new-born babies.

Accompanying rapid economic growth were significant structural changes. Table 5.5 lists the changing composition of the regional economy between 1980 and 1990. The general picture of structural change in this decade was a large proportional increase in manufacturing production and a simultaneous decline in the share of agricultural production. Clearly, agriculture as a traditional key economic sector is gradually giving way to manufacturing. Based on this agriculture to manufacturing transition, the process of industrialization appears to be taking place in the region.

Table 5.4

Economic growth for the Pearl River Delta, 1980-90

Items	Unit	1980	1990	% increase 1980-90
Per capita GVIAO[a]	yuan/person	113.34	5,488.09	+392.94
Employment rate	percent	50.70	58.70	+15.38
Rural per capita income[a]	yuan/person	238	1288	+441.18
Per capita export output	dollars/person	35.64	394.20	+100.06
Per capita realized foreign investment	dollars/person	5.74	75.49	+1,215.16

[a] Raw data are in 1980 constant prices.

Source: Guangdong, Statistical Bureau 1992b:65-7.

Rapid industrialization in the delta was not fuelled by the growth of large-scale, capital-intensive, modern manufacturing which is, as noted, difficult to develop in this region. The relative standing of heavy industry, which is mostly manufacturing of modern machinery, has declined since 1980 (see Table 5.5). The biggest gain in the share of total output value was made by the rural industrial sector, defined by the provincial authorities as industry located in the countryside (Guangdong, Statistical Bureau 1991a). Such rural industry is mostly small scale, labour intensive, and market oriented (Byrd and Lin 1990; Ho 1994; Lo 1989). Rural industry has since 1980 recorded not only the highest growth rate but also the biggest proportional increase in total industrial production (see Table 5.5). Rural industry appears to have become the most dynamic growth sector in the process of industrialization.

Alongside this industrialization, agriculture has become increasingly diversified and commercialized. While agricultural production has proportionally decreased, its absolute amount of the gross output value has continued to grow, although at a slower pace. Within the agricultural sector, food-grain production dropped from 75 percent to only 49 percent between 1980 and 1990. At the same time, engagement in market-oriented and high-profit industries such as forestry, livestock husbandry, sideline businesses, and fisheries grew at a faster speed, with a combined

Table 5.5

Changing composition of industrial and agricultural output value for the Pearl River Delta, 1980-90 (%)

Industry	1980	1990	Annual growth (%)[a]
GVIAO	100.00	100.00	19.24
Agriculture	38.60	19.33	6.86
Industry	61.40	80.67	21.23
Agriculture	100.00	100.00	6.86
Food grain	75.31	49.19	3.62
Forestry	1.82	2.98	8.89
Livestock husbandry	11.29	21.13	10.52
Sideline	6.28	14.05	12.66
Fishery	5.29	12.65	13.11
Industry	100.00	100.00	21.23
Light industry	63.23	64.91	22.05
Heavy industry	29.31	23.89	19.26
Rural industry	7.46	11.20	27.90

[a] Growth rates are calculated based on output value in 1980 constant prices.

Sources: Guangdong, Statistical Bureau 1991b:14-407; 1992b:65-6.

share of the total output value rising from 25 percent in 1980 to 50 percent in 1990 (see Table 5.5). Obviously, agriculture has been restructured from the traditional single-sided emphasis on food-grain production into a more diversified pattern, with commercial crops gaining an increasingly large share.

Spatial Redistribution of Economic Activities

Phenomenal economic growth and structural changes have brought about a process of spatial transformation in which economic activities, population, settlements, and land use are rearranged over space. Just as the growth of production does not occur evenly among various economic sectors, the magnitude of economic growth has varied significantly among cities and counties. The general picture of spatial change is characterized by the relative decline of Guangzhou as a dominant economic centre, and the accelerated growth of the counties and cities located in the areas between Guangzhou, Hong Kong, and Macao. Table 5.6 lists the growth of industrial and agricultural production in selected cities and counties in the delta region. It is not surprising that the two newly established Special Economic Zones of Shenzhen and Zhuhai have topped the whole region in the growth of industrial and agricultural production. Next to the SEZs come the newly developing counties and cities, including Baoan, Zhongshan, Dongguan, Shunde, Nanhai, Xinhui, and Panyu, which are mostly located in the Guangzhou-Hong Kong-Macao corridors. Medium-sized cities such as Huizhou, Foshan, and Zhaoqing have also continued to grow. By comparison, Guangzhou, the traditional primate city and chief economic centre of the region, recorded a low growth rate. Among the thirty-one counties and municipalities of the region, Guangzhou's production growth during the 1980s was the second lowest, higher only than that of Qingyuan shi, a mountainous underdeveloped municipality. The weakening of Guangzhou's dominant economic position in the region is clearly shown not only by its slower pace of economic growth but also by its share of the total regional production value, which dropped from 44 percent in 1980 to 22 percent in 1990 (see Table 5.6).

While the dominance of Guangzhou in the region has weakened, the Guangzhou-Hong Kong-Macao area has quickly emerged as a developing zone where production activities are concentrated. Map 5.2 shows the spatial distribution of the output value of the cities and counties of the delta on a comparable per capita basis. Clearly, industrial and agricultural production has been concentrated in the zones between Guangzhou, Hong Kong, and Macao. A similar pattern also existed in the spatial distribution of rural per capita income. Interestingly, higher rural per capita income occurred mostly around the central city of Guangzhou in 1980 (see Map

5.3). By 1990, the spatial pattern of distribution increasingly concentrated in the Guangzhou-Macao and Guangzhou-Hong Kong corridors (see Map 5.4). Although many counties and cities with a high rural per capita income were found in the Guangzhou-Macao corridor, they are in fact more closely connected with Hong Kong than Macao. Dongguan shi did not show a high rural per capita income, because since 1980 it has relied on compensational trade (*sanlai yibu*) as its major source of income. Such income, however, was counted separately by the local authorities as a processing fee (*gongjiaofei*) under the industrial category.

As compared with other regions in the country, the Pearl River Delta is distinguished by its remarkable performance in attracting foreign capital investment and promoting export production. The spatial distribution of realized foreign capital investment and export output value is displayed in Maps 5.5 and 5.6 respectively. Not surprisingly, the two Special Economic Zones of Shenzhen and Zhuhai are outstanding locales that lead the region in the acceptance of foreign capital and export of local products. The general pattern clearly demonstrates a concentration in the triangle bordered by Guangzhou, Hong Kong, and Macao. This pattern is consistent with the distribution pattern of industrial and agricultural output value, and the growth of per capita income (see Maps 5.2 and 5.4).

Table 5.6

Growth of GVIAO in selected cities and counties in the Pearl River Delta, 1980-90[a]

		Regional share	
Place	Annual growth %	1980	1990
Shenzhen	68.54	0.39	12.48
Zhuhai	39.31	0.69	3.27
Baoan	36.15	0.66	2.48
Huizhou	35.48	0.45	1.63
Zhongshan	23.28	4.73	6.60
Dongguan	23.10	4.54	6.24
Foshan	21.99	4.62	5.80
Shunde	20.96	5.06	5.83
Zhaoqing	20.66	1.21	1.37
Nanhai	19.79	5.09	5.33
Xinhui	19.74	2.98	3.11
Panyu	19.34	3.09	3.12
Jiangmen	17.14	3.33	2.79
Guangzhou	11.44	44.00	22.36

[a] Raw data are in 1980 constant prices.

Source: Guangdong, Statistical Bureau 1992b:83-206.

Does the declining degree of concentration in Guangzhou mean a decreasing spatial disparity of production and income for the whole region? Is the declining primacy of Guangzhou City an indication of the end of polarization and the beginning of a 'trickle-down' effect, as some writers have suggested? It is unfortunate that available data do not permit a systematic calculation of a Gini coefficient to show the changing spatial disparity. Nevertheless, data do allow a calculation of variance and standard deviation based on both the Gross Value of Industrial and Agricultural Output and the rural per capita income. The results show a significant increase in standard deviation for both per capita GVIAO and rural per capita income, with the former rising from 775 to 6,893 yuan and the latter increasing from 103 to 645 yuan between 1980 and 1990. This increase means that, despite the declining dominance of Guangzhou City in the region, spatial disparity in terms of per capita output value or rural per

Map 5.2

**Per capita industrial and agricultural output for the
Pearl River Delta, 1990**

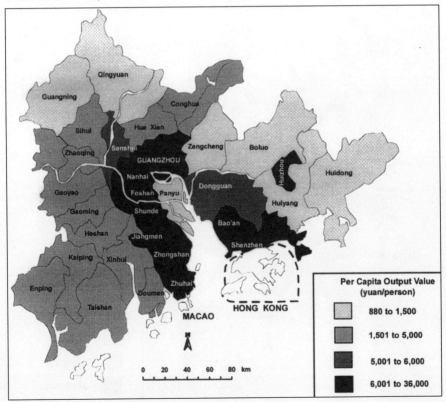

Source: Derived from Guangdong, Statistical Bureau 1991b:14-407.

capita income has remained considerably large. It has, in fact, widened throughout the 1980s. This finding is consistent with that of several recent studies which show that regional inequality in China since the reforms has been reduced at the interprovincial level but increased at the intraprovincial level (Fan 1995:443; Wei and Ma 1996).

Spatial Redistribution of Population and Migration

How and to what extent has the sectoral and spatial restructuring of economic activities affected the spatial redistribution of population? A remarkable spatial configuration of economic growth and structural change is the reinforced concentration of population in the Guangzhou-Hong Kong-Macao triangle zone. Maps 5.7 and 5.8 show the distribution of population density in the region for 1980 and 1990 respectively. Overall, population density for the entire region increased from 370 to 437 persons per square kilometre during the 1980s. As the natural increase of population

Map 5.3

Rural per capita income for the Pearl River Delta, 1980

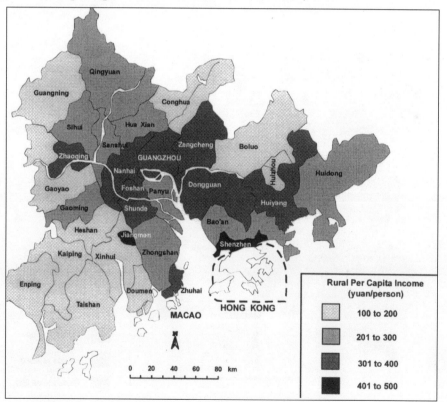

Source: Derived from Guangdong, Statistical Bureau 1991b:14-407.

was low because of the effective national campaign of family planning, any significant increase in population density must be primarily the result of migration from other areas.

When the spatial pattern of population density is examined, the higher population density in 1980 appears to have occurred mostly in the central delta, with the designated cities of Guangzhou, Foshan, Jiangmen, and Zhaoqing being the most populous places. This pattern carried on into the 1990s. It is no surprise that the Special Economic Zones of Shenzhen and Zhuhai have become much more populated than anywhere else in the delta due to their economic appeal. Shunde and Huizhou have also increased their ranks as a result of their new industrial developments. The two counties of Sanshui and Kaiping, which seemingly jumped to a higher rank in 1990, have not, in fact, changed very much. The pattern of population distribution has remained one of high concentration. The degree of spatial variation measured by the standard deviation of population den-

Map 5.4

Rural per capita income for the Pearl River Delta, 1990

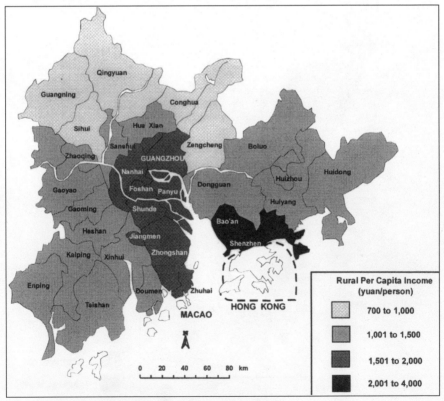

Source: Derived from Guangdong, Statistical Bureau 1991b:14-407.

sity increased from 719 in 1980 to 944 in 1990. Clearly, the spatial pattern of population distribution, which concentrated in the central delta at the expense of the periphery, was maintained and reinforced during the 1980s. This pattern resembles the one displayed by the distribution of industrial and agricultural production as identified above.

While the overall pattern of population concentration has remained virtually unchanged, except for a slight increase in regional variation, the mobility of the population has significantly increased as a result of both the process of economic restructuring and the recent relaxation of state control over population movement. As indicated, economic restructuring in the delta region was characterized by a relative decline in traditional farming and the rapid surge of rural industry. This process of economic restructuring significantly increased the potential mobility of the population in the region. On the one hand, the demise of traditional food-grain production released a sizeable number of rural labourers from the field. On

Map 5.5

Foreign capital investment in the Pearl River Delta, 1990

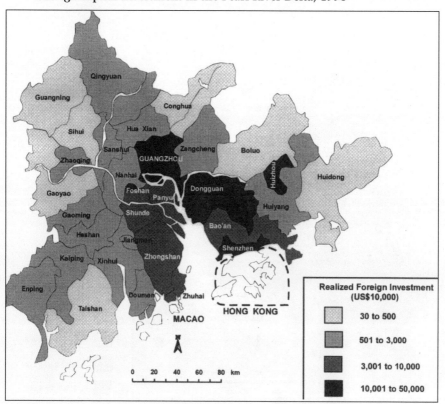

Source: Derived from Guangdong, Statistical Bureau 1991b:14-407.

the other hand, the flourishing of numerous rural industries, and the revitalized commercial activities in small towns, created phenomenal employment opportunities for the rural exodus. The combined effect of these push-and-pull forces has been a movement of people from farming to non-farming activities, and from rural to urban settlements. Such a movement has been facilitated since the mid-1980s by the state's deregulation of rural-urban migration.

The central state has recognized the growing number of surplus rural labourers that emerged as a result of economic restructuring in the countryside. In 1984, the State Council announced a new policy that permitted peasants to move to officially designated towns for settlement and to do non-agricultural jobs. In 1985, the state further relaxed its control on rural-urban migration. Under the new policy, peasants are allowed to move into nearby towns to establish stores, do construction work, or engage in transportation and other service jobs. They are treated like

Map 5.6

Per capita export value in the Pearl River Delta, 1990

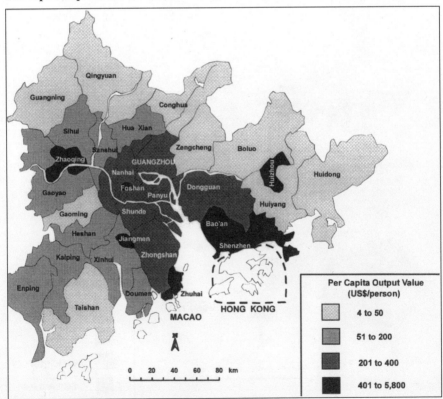

Source: Derived from Guangdong, Statistical Bureau 1991b:14-407.

other town residents, except that they must provide their own food grains without state subsidies (*zilikouliang*).

The restructuring of the regional economy combined with more liberal government policies for population movement resulted in extensive migrations to and within the delta. Before relevant data on migration are presented and analyzed, it is necessary to clarify the terminology and typology of migration in a Chinese context, because their misuse has been the source of much confusion. Basically, three major types of population movement take place in the delta. The first type of movement involves long-distance migration, where people move into the delta from other areas of the province or from other provinces. No data are available to show the exact number of these immigrants, but it is believed that they account for only a small proportion of the total migrants found in the delta (Xu and Li 1990:55-6). According to an official 1988 survey conducted by Guangdong Province, immigrants who originated from areas outside of Guangdong

Map 5.7

Population density for the Pearl River Delta, 1980

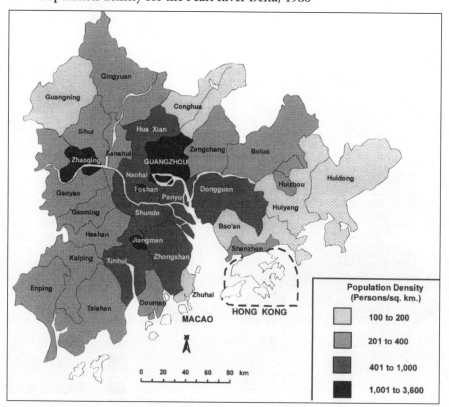

Source: Derived from Guangdong, Statistical Bureau 1991b:14-407.

accounted for only 11 percent of total migrants (Guangdong, Office for Population Census 1988:546-53). Among these newcomers, over 60 percent are unmarried young females (Guangdong, Office for Population Census 1988:546; Li 1989:43). Most of them work in factories and they speak various Chinese dialects other than Cantonese. It is probably because of their gender, occupational, and linguistic characteristics that the local people have commonly referred to them as 'the working girls' (*dagong mei*) or 'the girls from outside' (*wailai mei*). Statistically, these immigrants are counted as 'temporary population' (*zhanju renkou*), because their household registration status is still determined by their hometowns outside of Guangdong. This status is in spite of the fact that these immigrants have resided in the delta for one year or longer, and some intend to stay permanently.

The second type of population movement is the short-distance migration of local people within the delta region. Most of these local migrants are surplus rural labourers who have moved into nearby towns or small

Map 5.8

Population density for the Pearl River Delta, 1990

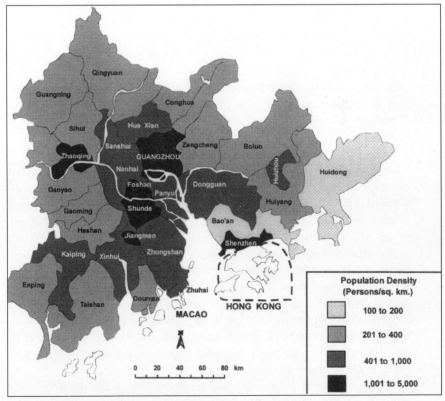

Source: Derived from Guangdong, Statistical Bureau 1991b:14-407.

cities for factory jobs, business activities, or construction work. These new townspeople may or may not obtain official urban registration status. Statistically, they are classified as 'residents who take care of their food grain' (*zili kouliang renkou*), or 'population who has not been registered as town residents' (*weiluohu changzhu renkou*). They were included in the category of 'temporary population' in the national census conducted in 1982 and in 1990. Available data are not systematic enough to show how many of them were in the delta region, but it was revealed in the 1988 population survey of Guangdong Province that relocated residents within the province accounted for 88 percent of total migrants (Guangdong, Office for Population Census 1988:546). A case study conducted by Xu and Li (1990:56) also revealed that migration in the Pearl River Delta region was mostly of a short distance in nature.

A third type of population movement involves short visits by tourists, government officials, businessmen, petty traders, and other people who come to the delta on a transient basis. These visitors are generally known as 'floating population' (*liudong renkou*). No systematic statistical record has been produced for this population. Surveys in individual cities such as Guangzhou and Shenzhen have indicated that in 1989 this group has reached as many as 1.3 million people daily in Guangzhou (Li and Hu 1991:8) and 300,000 in Shenzhen (Xu and Li 1990:55).

The three types of immigrants identified here have different demographic features and have different impacts on their destinations. Such differences have tended to be overlooked. Many research reports on migration in China have frequently confused the 'floating population' with 'temporary residents' or 'lodging population' and sometimes with 'residents who take care of their own food grains.'

Having clarified the terms and types of immigrants, it is now possible to assess the pattern of migration taking place in the delta. Data show that between the two national census years of 1982 and 1990, 'temporary population' (*zhanju renkou*), that is, a combination of the first and second types of migrants identified above, increased by 42 percent, or an average of 350,126 persons per year (see Table 5.7). This growth rate was much higher than the provincial average of 29 percent per year. The total number of the 'temporary population,' essentially migrants to and within the delta region, rose from 184,000 in 1982 to 2.98 million in 1990. The delta's share of the provincial total of temporary population jumped from 37 percent to 79 percent. In other words, almost 80 percent of the temporary population found in Guangdong Province in 1990 resided in the Pearl River Delta region. Obviously, the delta region has become the major destination for migrants in the province.

When analyzed at the intraregional level, the growth and distribution of migration have demonstrated a spatial pattern of concentration in the

Map 5.9

Annual growth of temporary population in the Pearl River Delta, 1982-90

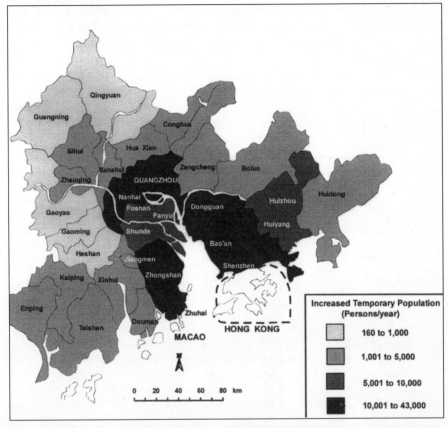

Source: Derived from Guangdong, Office for Population Census 1991:40-4.

Table 5.7

Temporary population in the Pearl River Delta and Guangdong Province, 1982-90

	1982	1990	Annual growth (%)
Pearl River Delta	183,778	2,984,788	41.69
Guangdong Province	493,413	3,791,002	29.03
Delta as % of Guangdong	37.25	78.73	

Source: Guangdong, Office for Population Census 1991:30-44.

central delta region. Map 5.9 depicts the spatial distribution of the annual growth of migration in the region between 1982 and 1990. Not surprisingly, the most dramatic growth of migration occurred in the two Special Economic Zones of Shenzhen and Zhuhai, both of which recorded an extraordinary annual growth rate of over 75 percent. Their share of the regional total of temporary population rose significantly from a mere 2.4 percent to 21 percent during the eight years between 1982 and 1990 (see Table 5.8).

The primate city of Guangzhou and other existing designated cities, such as Jiangmen, Zhaoqing, and Huizhou, did not receive many immigrants during this period. The growth rates of migration to these cities

Table 5.8

Changing distribution of temporary population in the Pearl River Delta, 1982-90

	1982		1990		Annual growth (%)
	Number	% of total	Number	% of total	
Guangzhou	70,541	**38.38**	451,761	**15.14**	**26.13**
SEZs[a]	4,438	**2.41**	640,593	**21.46**	**86.18**
Shenzhen	2,925	1.59	506,185	16.96	90.45
Zhuhai	1,513	0.82	134,408	4.50	75.22
Existing cities	23,312	**12.69**	217,996	**7.30**	**32.24**
Foshan	4,929	2.68	81,280	2.72	41.96
Jiangmen	3,325	1.81	22,827	0.76	27.23
Zhaoqing	8,724	4.75	43,837	1.47	22.36
Huizhou	6,334	3.45	70,052	2.35	35.04
Selected counties and cities	27,106	**14.75**	1,335,329	**44.74**	**62.77**
Zhongshan	6,368	3.47	108,884	3.65	42.60
Dongguan	5,228	2.84	453,005	15.18	74.67
Shunde	5,294	2.88	76,312	2.56	39.59
Nanhai	4,956	2.70	123,167	4.12	49.42
Panyu	2,904	1.58	44,795	1.50	40.78
Baoan	2,356	1.28	529,166	17.73	96.76
Other counties	58,381	**31.77**	339,109	**11.36**	**24.59**
Total	183,778	100.00	2,984,788	100.00	41.69

[a] SEZ stands for Special Economic Zone.

Source: Guangdong, Office for Population Census 1991:40-4.

were all lower than the regional average. Consequently, their share of the total temporary population of the region has dropped significantly (see Table 5.8). The only exception is Foshan, which had moderate growth and a slight proportional increase in temporary population. This increase is partly because of its intense economic linkage with the counties of the central delta, and partly because it contains the small town of Shiwan, which is relatively accessible to immigrants.

The most remarkable increase in temporary population occurred in the newly developing counties within the Guangzhou-Hong Kong-Macao triangle. As revealed in Table 5.8, these counties, especially Baoan and Dongguan, have experienced not only a much faster growth rate of in-migration than other parts of the region, but also a dramatic increase in their share of the regional total of temporary population. As a group, they tripled their share of the regional total in eight years, accounting for a disproportionate 45 percent of all temporary population in 1990.

While the SEZs and the newly developing counties were receiving an increasing and disproportionate number of migrants, the cities and counties on the periphery were left far behind. These peripheral cities and counties accounted for 72.97 percent of the delta's land area and 47.88 percent of its total population, but received only 31 percent of total immigrants in 1982. That disproportionately low percentage dropped even further to 11 percent in 1990 (see Table 5.8).

This spatial pattern of migration, characterized by a high concentration in the Guangzhou-Hong Kong-Macao triangle zone, by a large city whose primacy has declined, and by a relatively backward periphery is consistent with the spatial distribution of production activities identified here. While the proportional decline of migrants ending up in Guangzhou and other designated cities might have been the outcome of government limitation on migration to the cities, the tendency of migration to favour the central delta region over the periphery is very likely a result of the spontaneous process of economic growth and restructuring. According to a 1988 sample survey of 1 percent of the population conducted by Guangdong Province, the migration that occurred in the province in 1986-7 was mainly driven by the motive of seeking employment. Those migrants who moved for job-related reasons, such as transfers, new assignments, business start-ups, or factory work, accounted for about 70 percent of all sampled immigrants, more than any other category (see Table 5.9). Given that the Special Economic Zones and their surrounding counties can provide more job opportunities to the immigrants as a result of their rapid economic growth and restructuring, they have been favoured by more immigrants than places on the periphery. The corresponding relationship between the spatial distribution of the temporary population and that of production activities has also been revealed by a statistical analysis using

Pearson correlation coefficients that shows a strong positive relationship between the percentage of temporary population in each city or county and the percentage of employees (r = 0.64), per capita gross value of industrial and agricultural output (r = 0.73), and per capita income (r = 0.87). In other words, those counties or cities where there was a high proportion of temporary population tended to have high economic productivity and high per capita income.

The Growth and Distribution of Urban Population
How has the regional settlement system responded to economic growth, structural change, and increased population mobility? Did the rapid economic growth of the delta region result in a process of population concentration in the cities, as the conventional wisdom of urbanization might have expected? Before moving to answers for these questions, it is necessary to clarify the meaning of urban population and urbanization in the Chinese context.

The confusion over the definition of China's urban settlements and urban population has been discussed extensively (Chan and Xu 1985; Kirkby 1985; Ma and Cui 1987; Lee 1989). The total population of Chinese municipalities and towns is not recognized as an appropriate indicator for China's urban population, because that figure contains a large number of rural dwellers in the jurisdiction of the municipalities and towns, and therefore tends to exaggerate China's urbanization. In the case of the delta

Table 5.9

Reasons for migration in Guangdong Province, 1986-7

Reason	Number (thousands)	%
Job transfer	43.2	18.4
Job assignment	18.3	7.8
To enter business or do factory work	103.6	44.2
Education or job training	23.1	9.9
To live with friends or relatives	5.8	2.5
Retirement or job resignation	1.6	0.7
Dependants of migrants	10.0	4.3
Marriage	25.6	10.9
Other	3.1	1.3
Total	234.3	100.0

Note: Numbers refer to intra- and interprovincial migrants to towns in Guangdong, including those from villages as well as from cities and other towns but excluding temporary migrants who had lived in the destination for less than six months.

Source: China, State Statistical Bureau 1988:774-5.

region, the two newly designated shi or 'cities' of Dongguan and Zhongshan provide a good illustration. Each of these 'cities' had a total population in excess of one million, but, in reality, over 75 percent of their population was agricultural and less than 5 percent of their land area was truly built-up urban settlements. In Dongguan shi, the urban centre of Guangcheng had a population of only 120,000. Its built-up area in 1990 did not exceed 10 square kilometres. Zhongshan shi was in a similar situation. It would be inappropriate and misleading to take the total population of Dongguan and Zhongshan as their 'urban population' and herald them as two suddenly emerged 'extra-large cities' of more than one million people. The same problem also existed in Chinese designated towns. Under the new administrative system of 'town administering village' (*zhen guan cun*) implemented in the mid-1980s, most towns have incorporated or annexed extensive farming areas into their jurisdiction. These annexed rural areas automatically became zhen or townships, and their population, mostly peasantry, was automatically counted as 'town population' (*chengzhen renkou*).

In solving the problem of defining China's urban population, many scholars have suggested that the non-agricultural population of Chinese cities and towns be used as an indicator for China's de facto urban population (Ma and Cui 1987:389; Kirkby 1985:80; Lee 1989:774). This indicator appears to be more realistic and reliable than a count of the total population of shi and zhen. It does not, however, include those immigrants who have worked and resided in the shi and zhen but have not yet been granted the non-agricultural registration status for entitlement to government-subsidized food grain. Thus, counting only the non-agricultural population may sometimes underestimate the actual magnitude of China's urban population and urbanization. In some places such as Baoan and Dongguan, where a large temporary population lives in towns, the underestimation of the urban population is likely to be substantial.

The temporary population has, however, frequently moved in and out of towns on a transient basis. There are not yet any spatially systematic and historically comparable data to show how much of this temporary population has moved from the countryside to settle down permanently in towns and cities. It is, therefore, extremely difficult to assess the impact of the temporary population on the process of urbanization and settlement reorganization in the delta region.

For the purpose of data consistency and comparability, on both spatial and temporal bases, this study will use the non-agricultural population as a chief indicator of 'urbanness' to assess the rate and spatial distribution of urbanization in the delta region. Whenever possible, the proportion of the temporary population that has moved from the rural areas to reside for at least one year in the cities and towns will be estimated and the pattern of change analyzed to complement the non-agricultural measurement of 'urbanness.'

The growth of urbanization in the delta region in the 1980s was remarkable. Data have shown that the proportion of non-agricultural population for the entire region rose from 27.36 percent in 1980 to 36.71 percent in 1990 (Guangdong, Statistical Bureau 1992b:65). Such a calculation does not fully represent the true magnitude of urbanization because it does not include the temporary population that has resided in the cities and towns of the delta for a prolonged period of time. According to the official census data, the number of the total temporary population in the delta region increased from 183,778 in 1982 to 2,984,788 in 1990 (Guangdong, Office for Population Census 1991:40-4). But how many of those people have moved into cities and towns from the countryside within and outside Guangdong? There are no such data specifically for the Pearl River Delta region. The only way of finding out the impact of the temporary population on the delta's urbanization is to use a rough estimate. This estimate is made possible by the 1988 sample survey of 1 percent of the population conducted by Guangdong Province (Guangdong, Office for Population Census 1988:677, 706). According to this survey, of the total sampled migrants who moved into urban Guangdong from 1982 to 1987, 71.69 percent were originally from villages, and 97.59 percent of them had moved into the cities or towns in the province (Ma and Lin 1993:595). Assuming that these ratios of rural-urban migration are applicable to the delta region, an estimate can be made of the number of temporary individuals who originated in the rural areas and who have moved to urban settlements of cities or towns in the delta region (see Table 5.10). When the estimated urban temporary population is added to the non-agricultural population, the

Table 5.10

Growth of urban population in the Pearl River Delta, 1982-90

		1982	1990	Annual growth (%)
Total Population		18,104,670	20,801,194	1.75
Non-agricultural	Number	5,102,143	7,635,099	5.17
	%	28.18	36.71	
Urban temporary[a]	Number	128,576	1,496,833	35.91
	%	0.71	7.19	
Total urban	Number	5,230,719	9,131,932	7.21
	%	28.89	43.90	

[a] Numbers for urban temporary population are calculated according to the ratio derived from Table 5.13.

Sources: Guangdong, Statistical Bureau 1991b:14-407; Guangdong, Office for Population Census 1991:40-4.

delta region shows an urbanization level of 44 percent, which is significantly higher than both the 1990 national average of 26 percent or the provincial average of 36 percent. While this estimated urbanization level is not absolutely accurate, a 40 percent to 45 percent urbanization level for the delta region appears to be close to reality, given that the region includes a provincial capital with a population of about two million people, three prefecture-level cities, two Special Economic Zones, and numerous traditional small towns.

The data listed in Table 5.10 also indicate that the urban temporary population has been growing at a much faster pace than either the total or the non-agricultural population, although it still accounts for a very small proportion of the total population. This trend is because the temporary population is highly concentrated in Shenzhen, Baoan, and Dongguan. Therefore, the spatial impact they have on the whole region in terms of population concentration could be more significant than what is reflected by their share of the total population.

For the region as a whole, the pattern of urbanization seems to confirm the theoretical expectation that rapid economic growth will inevitably bring about an accelerated increase in urbanization. However, when the spatial distribution and the settlement composition of this increased urban population are examined in depth, some distinct features emerge that challenge conventional theoretical expectation. First, the spatial distribution of the percentage of urban population does not correspond very well with the pattern of distribution in production and migration identified earlier. Map 5.10 shows the spatial distribution of the percentage of non-agricultural population in the region in 1990. Except for the two Special Economic Zones and several traditional established cities, most places show a low non-agricultural proportion in their total population. This pattern is significantly different from the one on production and migration which showed a concentration in the triangle formed by Guangzhou, Hong Kong, and Macao. The non-agricultural indicator may be inadequate to reveal the spatial differentiation of urban population, because a considerable number of urban temporary residents do not have an officially granted non-agricultural registration status. In recognition of this possible inadequacy, the urban temporary residents were deliberately added to the non-agricultural population and their combination, called the urban population, is displayed in Map 5.11. The resulting spatial pattern still displays little change as compared to the previous map. What is intriguing is that those counties that have experienced the most dramatic economic growth in the central delta, such as Zhongshan, Shunde, Nanhai, Xinhui, and Panyu, remain predominantly agricultural/rural in their population composition. This pattern exists in spite of the fact that many of these places have achieved a considerably higher level of productivity and per capita income, and some of them have been recently designated as shi or cities by Chinese authorities.

Some important exceptions merit special attention. When the growth rate of the aggregate non-agricultural and urban temporary population is mapped, a distinct pattern emerges that shows a significant concentration of urban growth in the Shenzhen-Guangzhou corridor (see Map 5.12). The growth of the urban population in other counties in the central delta region remains relatively insignificant.

That the urbanization level and urban growth in the central delta region remain relatively low in spite of rapid economic growth in the area is to some extent not very surprising. Historically, this area has long been an agricultural region specializing in such activities as silviculture, aquaculture, and flower gardening, which are mostly labour intensive. Geographically, the central delta is divided by numerous tributaries and

Map 5.10

Percent non-agricultural population for the Pearl River Delta, 1990

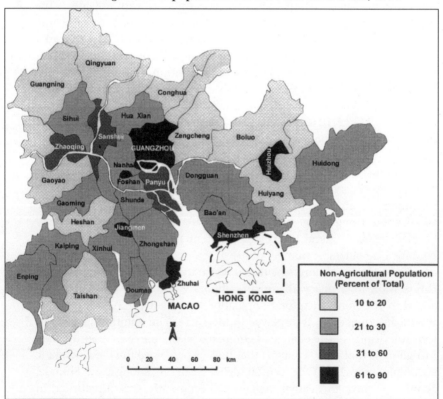

Source: Derived from Guangdong, Statistical Bureau 1991b:14-407.

studded with lofty hills. This geographical setting combined with a locally built landscape of a mulberry-dike-pond system has made it difficult to develop cities, which usually require a large continuous piece of land for the built-up area. During the years before the reforms, cities and towns in the delta had never experienced any substantial growth because of the declared Communist commitment of eliminating the rural-urban disparity, the suspicion that cities were vulnerable to nuclear attacks, and the prevailing economic policies that constantly assaulted urban commercial functions. From 1957 to 1978, urban population in the delta recorded a growth rate even lower than the national average (Xu and Li 1990:53). With such a low growth rate and its long agricultural tradition, it is not surprising that the central delta region has not yet reached a high level of urbanization.

The more fundamental force behind this low urbanization level in the

Map 5.11

Percent urban population for the Pearl River Delta, 1990

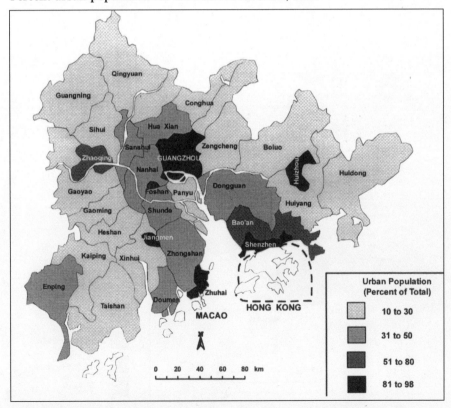

Source: Derived from Guangdong, Statistical Bureau 1991b:14-407; and Guangdong, Office for Population Census 1991:40-4.

central delta comes most likely from the area's special process of industrialization and economic restructuring. As discussed, economic restructuring in the delta region has been characterized by the rapid growth of small-scale and labour-intensive rural industry, which has been based in rural villages or small towns. The growth of numerous township and village industries since the 1980s has provided a great number of employment opportunities for surplus rural labourers. This spontaneous rural industrialization has been a fundamental force leading the rural exodus 'to leave the soil but not the village' (*litu bulixiang*) and 'to enter the factory but not the city' (*jinchang bujincheng*) (Ho 1994; Byrd and Lin 1990). In other words, the growth of rural industry has prevented surplus rural labour from flowing into the cities. This explanation can be supported by a statistical analysis using Pearson correlation coefficients that shows the percentage of the urban population has a significant negative correlation

Map 5.12

Annual growth of urban population in the Pearl River Delta, 1982-90

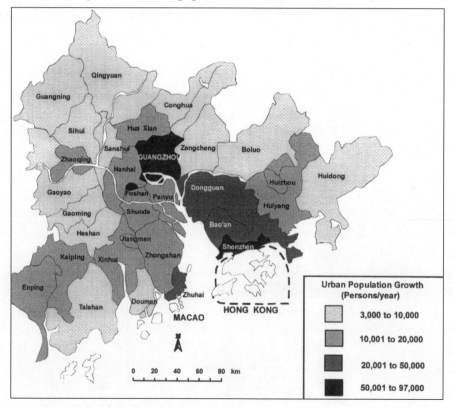

Source: Derived from Guangdong, Statistical Bureau 1991b:14-407; and Guangdong, Office for Population Census 1991:40-4.

with the percentage of rural industrial output value (r = -0.5996) and the percentage of rural labour in township and village enterprises (r = -0.5766).

The exceptionally high growth of the urban population in the Shenzhen-Guangzhou corridor was fuelled primarily by the inflow of investments from Hong Kong and overseas. Historically, this corridor, excluding Guangzhou, was underdeveloped for two reasons. First, its land resources were poorer than those in the central delta region, because most of the land was formed by recent fluvial and marine sedimentation. This land had a high salt content and was, therefore, unsuitable for intensive cultivation. Second, its frontier location near the capitalist territory of Hong Kong automatically excluded it from consideration in budget allocations made by the central state and the provincial governments. The open door policy implemented since 1979 has completely turned the locational disadvantage of this corridor into a rare geographical asset, a change bringing enormous economic prosperity. The corridor's proximity to Hong Kong and its autonomy in offering tax concessions and other preferential treatment to foreign investors have induced an unprecedented inflow of investments from Hong Kong and overseas. In 1990, the actual amount of realized foreign investment in the corridor of Shenzhen, Baoan, and Dongguan reached US$624 million, accounting for 41 percent of the total foreign investment in the whole delta region (Guangdong, Statistical Bureau 1991b:38, 42, 240). Numerous joint ventures and factories in the form of compensational trading have been set up not only in the Shenzhen Special Economic Zone but also in Baoan and Dongguan. The unprecedented development in this corridor has drawn a great number of labourers from all over the country. An estimated 1.488 million immigrants registered as temporary population were found in Shenzhen, Baoan, and Dongguan, which accounted for 50 percent of the total temporary population of the entire delta region (Guangdong, Office for Population Census 1991:40-4). Most of these temporary residents have been working as contract workers for joint ventures or other compensational trading enterprises in cities or towns. It is probably this temporary population that has driven the exceptional increase in urban population in the corridor.

In view of this spatial differentiation, we may surmise that two different patterns of urban growth exist in the delta. On the one hand, there is an emerging corridor between Guangzhou and Hong Kong, where urban population, particularly urban temporary residents, has been growing rapidly. On the other hand, there are areas outside the Guangzhou-Hong Kong corridor where the growth of urban population remains moderate in spite of remarkably rapid economic development. The former is primarily a spatial outcome of the intrusion of global market forces channelled through Hong Kong, whereas the latter is more likely driven by internal forces emanating from the local, spontaneous process of rural industrialization.

Reorganization of the Settlement System

The process of urbanization is manifest in the reorganization of the regional settlement system. An analysis of population growth according to different settlement types indicates that large cities in the delta region have not received much non-agricultural population. Instead, the numerous small towns widely scattered over the countryside have absorbed a large proportion of the newly emerged non-agricultural people, those who have changed their occupations from agricultural to non-agricultural pursuits and received official urban status (see Table 5.11). Smaller urban settlements have also become the most dynamic element in the region's settlement system, experiencing the fastest growth rate and gaining an increasing proportion of the total non-agricultural population. By comparison, the primate city of Guangzhou showed little expansion. Its share of the regional total has dropped from 53 percent in 1980 to only 34 percent in 1990 (see Table 5.12). This decrease is no surprise, as the economic production of the city has slowed down considerably during the past ten years. What is worthy of attention is the fact that the growth of cities other than Guangzhou has not been substantial. Most cities were in the small-city group, and the large-city category remained empty after ten years of rapid economic growth and development (see Table 5.12). The result of this analysis seems to suggest that cities, especially large cities, no longer set the pace for change in the delta's urbanization. Instead, numerous small towns emerging from the grassroots of the urban hierarchy have played an increasingly important role in fostering the process of urbanization and settlement transformation.

As the settlement system is shaped not only by the increase of the non-agricultural population but also by the influx of migrants, it would be inadequate to assess the process of settlement transformation without examining the movement of the temporary population. Unfortunately, no

Table 5.11

Changing non-agricultural population for the Pearl River Delta, 1980-90

Place	1980	1990	Increase	Share of total increase (%)
Primate city of Guangzhou	2,264,470	2,577,883	+313,413	9.77
Designated cities	580,172	1,603,086	+1,022,914	31.89
Designated towns	1,422,157	3,292,584	+1,871,427	58.34
Total	4,266,799	7,474,553	+3,207,754	100.00

Note: Data for designated cities do not include designated towns contained in their suburbs. They have been separately counted as designated towns.

Sources: Guangdong, Statistical Bureau 1986:2-41; 1990:108-30; 1991a:111-12; 1992b:83-206.

systematic data shows the movement of the temporary population among different types of settlements. Nevertheless, some interesting data do exist at the provincial level. According to a 1988 sample survey of 1 percent of the population conducted by Guangdong Province, during the five-year period from 1982 to 1987, cities, towns, and villages in the province received a total of 2.53 million migrants who had changed their residence officially, or who had left their place of origin more than six months earlier and had lived in the area for less than five years, regardless of their registration status. Of this total, about 71 percent ended up in small towns, 26.6 percent moved to cities, and the remaining less than 3 percent chose other villages (see Table 5.13). Obviously, among all types of settlements small towns have accommodated the largest proportion of immigrants. Among these immigrants, about 57 percent were female. In terms of their origin, 72 percent were from rural villages, which exceeds by a large margin the combined total of migrants originating from cities and towns. These figures do not prove that migrants were more interested in small

Table 5.12

Distribution of non-agricultural population among cities and towns in the Pearl River Delta, 1980-90

Cities and towns by size	No.	1980 Number of people	% of total	No.	1990 Number of people	% of total	Annual growth (%)
Extra large (>1,000,000)	1	2,264,470	53.07	1	2,577,883	34.49	1.30
Large (500,000-1,000,000)	0	0	0	0	0	0	
Medium (200,000-500,000)	0	0	0	2	604,780	8.09	
Small (100,000-200,000)	6	580,172	13.60	7	998,306	13.36	5.58
Towns (<100,000)	235	1,422,157	33.33	435	3,293,584	44.06	8.76
Total	242	4,266,799	100.00	445	7,474,553	100.00	5.77

Note: Data for cities do not include those towns contained in their suburbs as they have been separately counted in the category of towns. The three new designated cities in 1990 were Zhongshan, Dongguan, and Qingyuan.

Sources: Guangdong, Statistical Bureau 1986:2-41; 1990:108-30; 1991a:111-12; 1992b:83-206.

Table 5.13

**Population migration to and within
Guangdong Province, 1982-7**

	Number (thousands)	%
Total	2,535.6	100.00
Male	1,094.1	43.15
Female	1,441.4	56.85
Destinations		
Cities	698.5	27.55
Towns	1,776.0	70.04
Villages	61.1	2.41
Origin		
Cities	198.4	7.82
Towns	519.5	20.49
Villages	1,817.7	71.69

Source: China, State Statistical Bureau 1988:677, 706.

towns than in cities, given that it is normally much more difficult to obtain permission to move to cities than towns. However, the fact remains that under the prevailing policies governing population movement, the towns have unquestionably attracted more rural migrants than the cities. This finding is consistent with the result of a recent study conducted by Xu and Li, who reveal that small towns in the Pearl River Delta have attracted a larger number of rural surplus workers registering as 'lodging population' who supply their own food grain (*zili kouliang hu*). The primary reasons given for their preference for small towns are better job opportunities, improved availability of housing, and closer social ties to the migrants' villages of origin (Xu and Li 1990:55-6).

It is clear from the above analysis that, although the Pearl River Delta region has undoubtedly achieved rapid economic growth during the past decade, most places have remained predominantly agricultural or rural in their population composition. Even in the central delta region, where economic growth has been greatest, the population has not concentrated in the cities. Instead, small towns have played the leading role in this process of urbanization and settlement transformation. There is little evidence to show that rapid industrialization will necessarily result in a parallel movement of population toward the cities, as the conventional wisdom of urbanization has predicted. The deviation of the delta's experience of urban development from conventional expectations of urban transition results from the effects of both the government's relaxation of control over rural-town migration and the delta's self-sustained industrialization based primarily on the village and town. The unusually high urban growth in the Guangzhou-Hong Kong corridor, on the other hand, appears to be attributable to the operation of an external force, which has materialized in the form of capital investment from Hong Kong and overseas.

Land-Use Transformation

The rapid growth and restructuring of the delta's regional economy have also quickened the pace of land-use transformation. Industrialization, transport development, and the inflow of foreign investment have all created a great demand for land that can only be satisfied by reclaiming existing cultivated land. According to data released by Guangdong Province, between 1980 and 1990, a total of 327,800 acres of cultivated land was lost to non-agricultural uses in the delta region (Guangdong, Statistical Bureau 1992b:65). As a result, the total amount of existing cultivated land in the delta diminished from 2.58 million acres in 1980 to 2.25 million acres in 1990, and its per capita average dropped from 0.15 acre to 0.11 acres during this same period. This decrease in cultivated land represented an annual rate of -1.35 percent, almost double the provincial annual average of -0.85 percent. These figures come from an official source and do not include many unauthorized land-use transfers that have never been reported for tax reasons. In some places, such as Panyu, Dongguan, and Zhongshan, the actual amount of farmland loss has been much greater than was reflected in the official data. The problem of farmland disappearance was so severe that in 1985 the provincial government of Guangdong had to issue a directive limiting the annual amount of cultivated land that could be transferred for non-agricultural purposes. The government's statement, however, simply could not stop the process of cheap rice fields being converted into more profitable land tracts for the building of real estate, shopping malls, factories, or toll highways. It seems that local people were well aware that, to attract foreign capital investment, having cheap, unskilled labour is not enough, but providing cheap land with a good infrastructure is essential. Many cities, towns, and villages quickly reclaimed nearby agricultural land to form export-processing zones. Most of the developed land was also paved and, therefore, not reclaimable for agricultural use if the anticipated foreign investment failed to materialize.

Although the process of land-use transformation seems to be driven by the incentive of seeking higher profits and attracting more foreign investment, it is not known exactly how the lost cultivated land was redistributed according to industrial, transportation, commercial, and urban uses. No such data, at the local, regional, or national level, allow the construction of a balance sheet for land-use transfer. Nevertheless, by piecing together limited data from various sources, an analysis of general trends can still be made. Table 5.14 lists the changing urban and agricultural land uses for the delta region from 1980 to 1990. As expected, the substantial loss of cultivated land in the decade was accompanied by a simultaneous expansion of built-up areas in both cities and towns. Of the two types of urban settlement, designated cities showed relatively little expansion in their built-up areas. It should be noted that the 1990 figure includes the two newly designated 'cities' of Zhongshan and Qingyuan whose built-up areas were not included in the base figure for 1984. In other words, the expansion of the built-up area for these cities has been exaggerated because of administrative changes. Despite this exaggeration, the increase of built-up areas for cities remains relatively insignificant. On a comparable yearly basis, designated cities only expanded at an annual rate of 4 percent, much lower than either the expansion of built-up areas for designated towns or the reduction in cultivated land. It appears that the expansion of cities can provide little explanation for the drastic loss of the cultivated land. The annual expansion of the urban built-up area of designated cities covered only about 10 percent of the loss of cultivated land. This fact suggests that cities in the Pearl River Delta region did not experience significant expansion in land area.

The most significant expansion occurred in designated towns, which recorded the most rapid annual growth rate at 24 percent and covered about 50 percent of the 'lost' cultivated land when calculated on a comparable yearly basis. This dramatic expansion in designated towns was largely due to the relaxation of government criteria for town designation.

Table 5.14

Changing land uses in the Pearl River Delta (in acres), 1980-90

	1980	1984	1986	1990	Area of change (acres/yr)	Annual changing rate (%)
Cultivated land[1]	2,580,971			2,253,109	-32,786.2	-1.35
Built-up area of designated cities[2]		75,065.7		96,301.4	+3,539.3	+4.24
Built-up area of designated towns[3]	38,024.2		138,068.9		+16,674.1	+23.98

Sources:
[1] Guangdong, Statistical Bureau 1992b:65.
[2] China, State Statistical Bureau 1985:48 and China, State Statistical Bureau, 1991b:72.
[3] Guangdong, Statistical Bureau 1986:502-41.

Between 1980 and 1986, a total of 214 new towns were officially approved as designated towns. They covered a total built-up area of 158.34 square kilometres or 39,098.346 acres. When these reclassified built-up areas were deducted from the calculation, the net expansion of the existing towns still amounted to 60,946.398 acres during the six years from 1980 to 1986. On a yearly basis, existing small towns had been expanding at 10,157.733 acres, over three times the average annual expansion of designated cities. In a manner similar to population growth, small towns have been the most dynamic urban element, experiencing the fastest expansion in land.

A more interesting aspect of the delta's land-use transformation is the rapid expansion of non-agricultural land outside cities and towns. As shown in Table 5.14, the combined areal expansion of both cities and towns did not cover the total loss of cultivated land when calculated on a comparable annual basis. Urban sprawl from both cities and towns accounted for about 60 percent of farmland loss. A large portion of urban expansion was due to the official designation of two new cities and 214 new towns during this period. Strictly speaking, the built-up areas of these new cities and towns should not be taken into account, because they already existed and had little to do with the loss of the cultivated land under study. When these existing reclassified cities and towns were deducted, the net gain of urban built-up area for both cities and towns accounted for only about 42 percent of the annual loss of cultivated land. The remaining 58 percent of farmland loss must be claimed by non-agricultural uses taking place in the countryside other than the expansion of cities and towns.

The portion of farmland loss in the countryside has been substantial, with an average annual decrease of 19,091 acres. In other words, the Pearl River Delta region loses 77.32 square kilometres of its cultivated land to non-agricultural uses in the countryside every year. Such non-agricultural use could involve the construction of highways, development of real estate, establishment of industrial zones, or creation of a strip of shopping stores such as those in Shunde or Panyu. These structures may differ from one another in their uses, but they have shared a common feature: they are all located in the countryside. Consequently, it is not uncommon to find a few factories or a group of villa-style apartment buildings standing right in the middle of a rice field. Most of these factories and housing units are within walking distance and are easily accessible to the local peasants who have 'left the soil but not the village' or 'entered the factories but not the city.' The special process of rapid rural industrialization without the concentration of population in cities has thus created a unique land-use pattern characterized by a mix of intensive industrial/agricultural or urban/rural activities.

The fact that the largest proportion of farmland loss was taken for non-agricultural use in the countryside rather than encroached upon by built-up areas of the cities suggests that the driving force for the delta's land-use transformation is the rapid industrialization of the countryside, not the expansion of the existing cities. This observation has been confirmed by a statistical analysis that shows that the decreasing annual rate of cultivated land in the delta region has significant correlations with the annual growth rate of per capita industrial and agricultural output value ($r = 0.7183$). Interestingly, the decreasing rate of cultivated land has also shown a fairly significant correlation with the distance from Hong Kong ($r = 0.5060$) but no significant relationship with the distance from Guangzhou. It has high correlations with per capita transport investment ($r = 0.8602$) and per capita export output value ($r = 0.9488$). These results seem to suggest that transport development and the influence of external forces from Hong Kong are important factors responsible for the reduction of cultivated land.

As discussed, much farmland has been turned over for infrastructure development and for export-processing zones. This finding can also be illustrated with a map showing the spatial differentiation of the decreasing annual rate of cultivated land (see Map 5.13). The most severe loss of cultivated land was found in Dongguan, Baoan, and Zhongshan. These places were all characterized by their intense connections with Hong Kong and they have played a leading role in the delta's export production. The two Special Economic Zones of Shenzhen and Zhuhai did not show a high farmland loss, because the areas they occupied were smaller and hence the acreage lost was relatively insignificant. Doumen xian is the only county that has gained more cultivated land due to land reclamation from the sea. In any event, this analysis suggests that rural industrialization, transport development, and export-goods processing are likely the three important factors underlying the process of land-use transformation in the delta region. This finding reinforces those derived from the above analyses of the growth and distribution of the urban population.

The Emerging Spatial Pattern: A Statistical Analysis

This discussion has identified a number of spatial patterns related to the distribution of production, population, migration, urbanization, and land-use transformation. Although these patterns are not all identical because they represent different variables, they have displayed some similar features and presented consistent findings. To better describe the general geographical pattern of growth and development, the statistical methods of principal components and cluster analyses were employed to extract the common elements of distribution in development variables and to classify locations according to the similarities in extracted components.

All thirty-one cities (*shiqu*) and counties in the region were used as cases for the statistical analysis. The selection of variables was limited by the availability of data. Nevertheless, eight variables were selected, after considering the following factors. First, the variables selected must reflect the spatial distribution of production, population, and land use on a comparable basis. Second, to identify the resulting spatial differentiation of development, the variables selected must pertain to the spatial outcome of development. Finally, the number of variables selected should not exceed the number of cases and should allow the necessary degrees of freedom for the statistical computation of the interrelationship among the variables.

The spatial differentiation of production is denoted by per capita gross value of industrial and agricultural output (PGVIAO). As export production is a distinctive element of the economic growth of the delta, per capita export output value (PEXPT) is included as a variable for the analysis. Another two economic variables denote employment rate and per capita income (respectively, EMPY and PICM). The spatial variation in population, urbanization, and migration is represented by the variables of population density (DNTY), percent urban population (URBN), and per-

Map 5.13

Annual loss of cultivated land in the Pearl River Delta, 1980-90

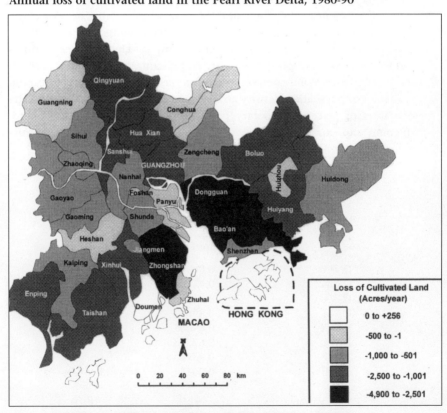

Source: Derived from Guangdong, Statistical Bureau 1991b:14-407.

cent temporary population (TEMP). The selection of variables on land use is hampered by the lack of data. The only available data pertaining to land use are the amount of cultivated land, the main resource of the delta. Per capita cultivated land (PCUL) is, therefore, chosen as a variable to reflect the spatial characteristics of land use. Thus, the eight selected variables represent the spatial differentiation in production, employment, income, population, urbanization, migration, and land use.

The statistical procedure begins with a principal components analysis, a common technique used to identify the fundamental dimensions of variations hidden beneath the complex surface of an area. Such a statistical technique enables many interrelated and complicated phenomena to be represented and described by a small number of principal components.

Two principal components are extracted as the main dimensions underlying the spatial pattern of development in the eight selected variables. They account for 80.6 percent of the total variance and are adequate to represent the spatial differentiation of the delta in production, population, and land use. A close examination of the rotated factor matrix (see Table 5.15) suggests that the first extracted component is an indication of the development in the newly developed area of the delta, because it has high factor loadings on the variables of percent temporary population, and per capita industrial and agricultural output value, but low loadings on the variables of population density and urbanization. It stands in sharp contrast to the second component, which has high factor loadings on the variables of population density and urbanization but low loadings on the variables of percent temporary population, per capita cultivated land, income, and export output value. Clearly, the first component represents

Table 5.15

Rotated factor matrix from principal components analysis

Variable	Factor Loadings	
	Factor One	Factor Two
DNTY	0.06099	0.91267
URBN	0.46426	0.81925
PGVIAO	0.81567	0.49424
PICM	0.89607	0.12471
PEXPT	0.89045	0.25589
EMPY	0.67223	0.29989
TEMP	0.93166	0.06726
PCUL	-0.20431	-0.82587
% of variance	61.90	18.70

Source: Computed from Guangdong, Statistical Bureau 1991b:14-407.

the element of new development that has taken place mostly outside the cities, whereas the second component describes the development features of the traditional cities where population density and percent urban population remain high but the percentage of temporary population is low, possibly because of the state's continued restriction on migration to the cities. Thus, the result of principal components analysis highlights the distinctive feature of development in the delta region, where higher production, export, income and in-migration on a comparable ratio or per capita basis did not occur in the existing large cities but in the areas where urbanization remained relatively low.

A cluster analysis was then performed to classify various geographical units according to their similarities in the loadings of the principal components. This analysis presents general pattern of spatial distribution in production, population, and land use in the delta region. The result of the cluster analysis is shown in Map 5.14.

Map 5.14

Result of cluster analysis for the Pearl River Delta

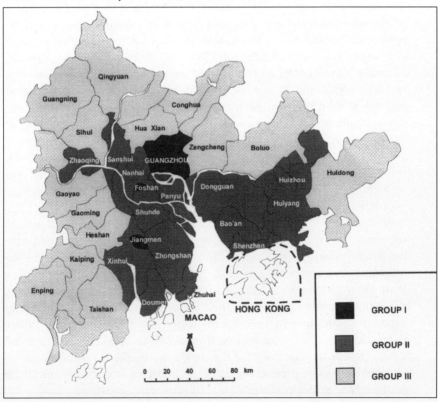

Based on their characteristics of development in production, population, and land use, the cities and counties of the delta region are classified into three groups. Group I consists of the cities of Guangzhou, Foshan, and Jiangmen, which are all traditional urban centres in the region. Group II includes all places located within the triangle area formed by Guangzhou, Hong Kong, and Macao. It is in this newly developing zone that rapid growth has been taking place since the 1980s. The cities of Huizhou and Zhaoqing are included in Group II probably because they show a development pattern more similar to the newly developing zone than to the traditional city type. Group III covers all places on the periphery of the region. The general pattern of spatial development of the delta region can thus be deconstructed into three tiers: traditional urban centres, the newly developing zone, and the periphery area.

How do these three groups differ from one another in terms of the level and pace of development? Tables 5.16A and 5.16B list some of the key economic and demographic indicators for the three groups. It appears that the most significant economic growth occurred in Group II, that is, the newly developing zone lying between or adjacent to major metropolitan centres. This group accounted for 35 percent of the delta's total land area and 40 percent of its total population in 1990, but it produced 55 percent of the total industrial and agricultural output of the delta region and 70 percent of its total export revenue. It has also become the chief recipient of foreign investment and the most favoured destination for immigrants, receiving 70 percent of all the realized amount of foreign investment in the delta region and 74 percent of the delta's total temporary population in 1990. More striking is its extraordinarily high growth rates of industrial and agricultural output, income, immigration, and foreign investment, all of which have exceeded both the traditional city group and the periphery area (see Tables 5.16A and 5.16B). The result of such accelerated growth was a dramatic increase in the 1980s of its regional share of industrial and agricultural output, export revenue, foreign investment, and temporary population.

While the newly developing zone was experiencing the most dramatic economic growth, the traditional urban centres in Group I exhibited only moderate growth in production and population. The growth rates in industrial and agricultural output, income, and foreign investment of these urban centres on a per capita basis were all significantly lower than the regional average. Moreover, their regional share in production value, export revenue, foreign investment, temporary population, and urban population dropped dramatically from 1980 to 1990. This pattern of urban growth has contradicted the neoclassic theoretical expectation of polarization, where production and population are said to concentrate in major urban centres at an early stage of economic growth.

Table 5.16A

Key economic and demographic indicators for the three zones grouped by cluster analysis for the Pearl River Delta

		Land (km²)	Total population 1980	Total population 1990	GVIAO[e] 1980	GVIAO[e] 1990	Temporary population 1982	Temporary population 1990
Group I[a]	No.	1,650	3,512,639	4,231,088	10,176.35	35,236.11	78,795	555,868
	%[d]	3.48	19.97	20.34	51.95	30.96	42.87	18.62
Group II[b]	No.	11,831	7,008,802	8,526,929	6,224.64	62,281.14	61,530	2,216,279
	%[d]	35.49	39.83	40.99	31.78	54.72	33.49	74.26
Group III[c]	No.	28,950	7,072,499	8,043,177	3,187.57	16,310.69	43,454	212,641
	%[d]	61.03	40.20	38.67	16.27	14.32	23.64	7.12
Total	No.	47,431	17,593,940	20,801,194	19,588.56	113,827.94	183,779	2,984,788
	%[d]	100.00	100.00	100.00	100.00	100.00	100.00	100.00

Notes:
[a] Group I includes traditional urban centres such as Guangzhou, Foshan, and Jiangmen.
[b] Group II includes those counties and cities located primarily in the triangle area between Hong Kong, Guangzhou, and Macao.
[c] Group III includes counties in the periphery zone of the delta region.
[d] Percentages refer to the share of the total for the whole delta region.
[e] Gross Value of Industrial and Agricultural Output in million yuan. Data are in 1980 constant prices.

Sources: Guangdong, Statistical Bureau 1991b:14-407; Guangdong, Office for Population Census 1991:40-4.

Table 5.16B

Key economic and demographic indicators for the three zones grouped by cluster analysis for the Pearl River Delta

		Urban population		Foreign investment (US $10,000)		Export output value (US $10,000)	
		1982	1990	1980	1990	1980	1990
Group I[a]	No.	2,788,911	3,799,432	3,449	32,671	35,028	180,076
	%[d]	53.32	41.61	34.08	20.96	55.14	24.55
Group II[b]	No.	1,597,112	3,692,907	6,332	109,436	26,536	512,182
	%[d]	30.53	40.44	62.56	70.23	41.78	69.83
Group III[c]	No.	844,695	1,639,592	340	13,729	1,957	41,231
	%[d]	16.15	17.95	3.36	8.81	3.08	5.62
Total	No.	5,230,719	9,131,932	10,121	155,836	63,521	733,489
	%[d]	100.00	100.00	100.00	100.00	100.00	100.00

Notes:
[a] Group I includes traditional urban centres such as Guangzhou, Foshan, and Jiangmen.
[b] Group II includes those counties and cities located primarily in the triangle area between Hong Kong, Guangzhou, and Macao.
[c] Group III includes counties in the periphery zone of the delta region.
[d] Percentages refer to the share of the total for the whole delta region.

Sources: Guangdong, Statistical Bureau 1991b:14-407; Guangdong, Office for Population Census 1991: 40-4.

Another important spatial aspect of development revealed by the statistical analysis is the relative underdevelopment of the periphery area (Group III). Despite the fact that the periphery area accounted for 61 percent of the delta's land area and 38 percent of its total population, the periphery contributed only 14 percent of industrial and agricultural outputs, 5 percent of export revenue, 8 percent of foreign investment, and 7 percent of the total temporary population to the region. Its regional contributions in industrial and agricultural production and acceptance of temporary population, which were already disproportionately low, dropped even further during the 1980s. Consequently, the disparity between the periphery area and the other two developed groups in per capita income and per capita output production remained large. There is little evidence to suggest that the 'trickle-down' effect has prevailed in the delta and that its regional disparity has been reduced. The issue of regional inequality in post-reform China appears to be more complicated than what the conventional wisdom of polarization might have predicted. While spatial inequality at the interprovincial level has been reduced because of the economic upsurge of the South China region, intraprovincial disparity has remained large and has even been widened.

It has become clear that the general pattern of spatial development in the delta region is characterized by the declining dominance of the traditional large cities, the rapid surge of a newly developing zone lying between major metropolitan centres, and the relative underdevelopment of the periphery area. This distinctive pattern identified by the statistical analysis is consistent with the qualitative analysis presented in previous sections of this chapter.

Summary

Rapid economic growth has been taking place in the Pearl River Delta region since economic reforms and the open door policy were initiated in the late 1970s. In a manner similar to development in other parts of China, the economic surge in the delta region has been primarily driven by township and village industries, which have emerged from the grassroots. Such industries are mostly rural based, small scale, unsophisticated, and labour intensive, but they have created far-reaching impacts on economic restructuring, town development, and land-use transformation. The growth and restructuring of the delta's economy are distinctive in that foreign investment and export production have played a significant part in the process. It is the dialectical interaction of these local and global forces that has brought profound changes to the delta's spatial economy.

The spatial configuration of the growth and restructuring of the delta's regional economy, as revealed by both quantitative and qualitative analyses, is characterized by an increasing concentration of production and population in the triangle area formed by Guangzhou, Hong Kong, and Macao. The primate city of Guangzhou has not exhibited any accelerated growth in production, population, or land area. It is the area lying between Guangzhou, Hong Kong, and Macao that has rapidly increased its growth rate and regional share of production and population.

Economic development in the delta has also resulted in a process of settlement reorganization. While the growth and restructuring of the regional economy has significantly fostered migration to and within the region, the population has not concentrated in the cities, especially the large cities. Instead, numerous small towns located mostly in the areas between or immediately adjacent to the large urban centres have accepted a growing amount of surplus rural labour.

Similarly, the expansion of cities has not contributed much to the transformation of land use. The magnitude of change in the countryside has been much greater than in the cities. Instead of forming new cities and towns at the cost of the existing agricultural land, the land-use transformation of the delta has been characterized by a redistribution of land within the countryside, resulting in a growing mix of industrial/agricultural or urban/rural activities.

How can we explain this distinct pattern of economic and spatial transformation? Preliminary conceptual and statistical analyses have suggested that rural industrialization and the intrusion of global market forces through Hong Kong are two noticeable factors underlying the process of economic and spatial transformation. But the detailed cause-effect relationship is far from clear. Are these factors the only ones that have shaped the economic and spatial development of the delta region? How and to what extent have these and other possible factors changed the development landscape? What are the exact roles of these factors in the process of transformation? How do they act individually and interact with one another in creating this distinctive pattern of growth and development? What are the implications of such interaction for our understanding of local-global dialectics in the process of regional development? Answering these questions requires an in-depth assessment of different individual cases.

In what follows, detailed case studies are conducted to investigate the operating mechanism of economic and spatial transformation. Although regional development is a complex outcome of social, economic, historical, and geographical factors, it is not possible to investigate every facet of the process. A manageable alternative is to focus the investigation on those key factors that have played the most significant role in the process of economic and spatial transformation. The identification of the 'most important' factors is not, however, an easy task, because it is a highly subjective matter reflecting different individual perspectives.

The following chapters examine the transformation of three selected counties where the impacts of rural industrialization, global forces, and transport infrastructure development are most noticeable. The cases and the key factors with which they are associated were selected after extensive consultation with experienced local Chinese researchers and my own field investigation. The selection will be justified at the beginning of each chapter.

The places and factors selected for these case studies may not completely represent the growth of the entire delta region, where there has been great geographical variation in development conditions. However, by focusing on some of the most dynamic areas, where the impacts of local and global forces are remarkable, the selected case studies may generate significant insights into the operating mechanism of development in the social, political, and economic context of the Pearl River Delta region.

6
Rural Industrialization

This chapter examines how a rural economy in the Pearl River Delta was transformed after economic reforms were initiated in 1978. Through a detailed case study, the chapter attempts to unfold the complex process of agricultural restructuring and rural industrialization that has occurred in the countryside, and to seek an explanation for the distinct features of spatial transformation in the delta region identified in Chapter 5. The importance of agricultural production has been fundamental to the regional economy of the Pearl River Delta. Until recently, modern manufacturing in this region had never been fully developed and its role in national industrial production was relatively insignificant when compared with other regions in the northeast and the north, particularly in the provinces of Liaoning, Shanxi, and Hubei. However, the delta has been outstanding in national agricultural production. Its excellent natural endowment plus its locational advantage for marketing farm produce overseas has enabled the region to become the major rice bowl of South China and the nation's chief exporter of many agricultural products, including pond fish, vegetables, and fruit. During the past four decades of Communist rule, 'agriculture the base' was highly lauded all over the nation, not least in the delta. When economic reforms were initiated in 1978, they started with agriculture. Today, after more than ten years of reform and restructuring, agriculture remains a vital sector of the delta's regional economy. Without a thorough investigation of changes in the agricultural sector, it would be impossible to understand the process of economic and spatial transformation.

Rural industrialization is also a crucial element of post-reform development in the countryside. As identified in Chapter 5, the most remarkable growth in production and employment occurred in the sector frequently referred to as 'rural industry' (*nongcun gongye*) or 'township and village industry' (*xiangzhen gongye*). The restructuring of the whole regional economy owes a great deal to a rapid surge in rural industry, which underscores many distinctive features in population redistribution, migration, and settlement reorganization. In many respects, rural industrialization has been one of the most dynamic elements of change, requiring an in-depth assessment to comprehend the dramatic transformation of the regional economy.

Nanhai xian, previously a rural county but now an industrialized municipality, is a typical case of the transformation of a rural economy after the implementation of economic reforms. With an area of 1,152 square kilometres and a population of over one million in 1994, Nanhai is, in many respects, a good representation of the Pearl River Delta region on a smaller scale. Geographically, Nanhai is composed of three main areas. The southern half of its land extends to the lower reach of the Pearl River and is dominated by a built agricultural system of fish ponds and mulberry dikes, a system popular for decades in many areas of the delta (Zhong 1982:191-202). Its central and eastern part is primarily flat plain suitable for the cultivation of paddy rice, vegetables, and fruit. Located on the immediate outskirts of the cities of Guangzhou and Foshan (see Map 1.2), this area serves as a suburb for the two cities, producing many farm products for urban consumption. The northern part extends to the upper reach of the North River (*Beijiang*), where the landscape is relatively mountainous and agricultural production is dominated by groundnut, sugar cane, and livestock. From a geographical point of view, Nanhai is diversified enough to illustrate the general situation of the Pearl River Delta.

As one of the 'four little tigers' (that is, the four county-level economies of Zhongshan, Dongguan, Shunde, and Nanhai; see Zhao 1991) that have recently emerged in the Pearl River Delta, Nanhai is renowned mostly for its successful agricultural restructuring and rapid industrialization. Specifically, Nanhai is outstanding in three achievements: rapid diversification and specialization in agriculture, a flourishing rural industry at the local level, and successful mechanization of the cultivation of paddy rice. Its most well-known feature is a comprehensive approach to rural industrialization, in which the maximum development is achieved simultaneously at five levels: county, township, district, village, and individual. The Nanhai Model, known as 'driving forward on five wheels' (*wuge lunzi yiqi zhuan*), has now been recognized by many Chinese planners and scholars as one of China's three national models for rural development in the post-reform era, standing parallel to, but distinct from, the Sunan Model, which emphasizes the collective sector, and the Wenzhou Model, whose mainstay is the private sector (Zhao 1991:159).

Historical Background

When economic reforms were initiated at the end of 1978, Nanhai was an overwhelmingly rural county, with three-quarters of its population earning a living from agriculture without the benefit of mechanization. The county was densely populated at 768 persons per square kilometre, which exceeded the provincial average by 468 and the national average by 658. Available farmland was so limited that cultivated land per capita was a mere 0.7 mu, significantly less than the provincial or national average. With limited farmland bearing a dense and growing population, underemployment was a long-standing problem. It has been estimated that on the eve of the reforms, about 40 percent of Nanhai's agricultural labour force was unable to find jobs (Zhao 1988:355). As rural-urban migration was restricted, the surplus rural labour had to remain on the farms for subsistence. The output of agricultural production did expand, but because of the large and growing population, labour productivity was low and little improvement occurred. The value of net output produced by farmers was a mere 549 yuan per labourer and rural per capita income was only 186 yuan in 1978. This situation, in which agriculture stagnates at a subsistence level, appears to fit neatly into the concept of 'involutionary growth' or 'growth without development,' which refers to the process of agricultural production expanding but marginal returns diminishing (Huang 1990:13).

At the time, local people were preoccupied with the task of seeking self-reliance in food supply. Grain production was taken as the key link in the economy and little room was available for other economic pursuits, including sideline farming, trading, transportation, and manufacturing. Such a single-sided production system, which overemphasized the rice crop, produced little material for manufacturing. An agrarian economy, perpetuated at the subsistence level, generated no surplus for capital investment in the manufacturing sector. Moreover, Nanhai was financially under such rigid control by both the provincial government and the Foshan City, to which it was subordinate, that it was unable to initiate industrial development at the local level. Consequently, modern manufacturing was virtually non-existent, except for a few small rural workshops repairing or producing some items for agricultural production, including fertilizer, cement, farm tools and machinery, and irrigation pumps.

Before the reforms, urban development had suffered severely from a constant attack on urban commercial functions, a lack of funds for town construction, and the government's restriction on rural-urban migration. The presence of Foshan City (*shiqu*) nearby had also taken away from Nanhai most urban functions along with construction funds. As a result, Nanhai in the pre-reform period recorded an abnormally low level of urbanization, with less than a quarter of its population engaged in non-agricultural activities that took place mainly in rural market towns. Nanhai did not even have a county town in the years before the reforms. It was not until 1990 that a county capital was built in the market town of Guicheng. The town was quickly developed to acquire designated city status, which was granted directly by the State Council in 1992. Thus, despite its central location in the delta and excellent natural endowments, Nanhai had been unable to ignite the engine of economic development until the 1980s.

Economic Reforms

The reform programs initiated at the Third Plenum of the 11th Central Committee of the Chinese Communist Party in December 1978 brought fundamental changes to the local economy. Although many factors are responsible for the development of the local economy, from an institutional point of view, four sets of policy changes underscored the process of economic transformation.

First, the implementation of the Household Production Responsibility System (HPRS) has provided great incentives for farmers to produce more for a higher profit. The HPRS, although existing in various forms and modified over time, is essentially a system of de-collectivization that decentralizes production decision-making from collective units (commune-brigade-team) to peasant households. The most popular form of the HPRS is 'contracts output to the household' (*baochan daohu*). Under such a system, a peasant household contracts with the production team to secure a piece of land or a number of fish ponds for fifteen years or longer. While the contracted land or fish ponds remain collectively owned, the household can decide on the crops or the types of fish, provided that it agrees to fulfil the contracted output quotas for both the state and the collective unit. Anything produced above the quotas can be sold either to the state at an above-quota price, or in the local free market at a negotiated price. The resulting profits can be retained by the household. In this way, the effort put into farm production is directly linked to profits. Those who are willing to work harder will be rewarded. This system signifies a drastic departure from the Maoist approach of egalitarianism. It has provided great incentives for farmers and invigorated the local economy.

Second, in an effort to raise farm income, the procurement prices for many farm products have increased sharply. Beginning with the 1979 summer harvest, grain quota procurement prices were raised by 20 percent, with an additional 50 percent premium for above-quota sales. The purchase prices of cotton, oil-bearing crops, sugar cane, pigs, cattle, eggs, and other farm and sideline products were also raised by an average of 22 percent. At the same time, the cost of agricultural input was reduced by 10 percent to 15 percent in 1979-80 (Ash 1988:540; Walker 1984: Riskin 1987:285). Such price adjustments apparently favour peasants and have further motivated farmers to increase agricultural production for higher income.

Third, the opening up of commodity markets has facilitated the transfer of resources in and out of the agricultural sector, and accelerated agricultural specialization. In the pre-reform years, Nanhai was, like other places in China, dominated by a marketing system under which the purchase and sale of farm products was a state monopoly (*tongguo tongxiao*). Agricultural production was carried out according to the orders of the state, and all farm products were purchased by the state at quota prices. Under this system, farmers had little knowledge of actual market demand. As grain production was universally taken as the key link in agriculture, farmers could not specialize in other activities that were more profitable or suitable for local conditions.

This state monopoly purchase and marketing system, which lasted for thirty years, was fundamentally reformed in 1985 to give free market forces a much larger role to play. The number of agricultural products that used to be purchased by the state was reduced from 118 to only 5 (Zeng 1988:2-3), and manufactured products supplied and distributed by the state to peasants from 95 to 13. Before the reforms, the provision of daily necessities such as grain, cooking oil, meat, fish, sugar, cloth, and soap was monopolized by the state through a rationing system under which consumers could not buy what they needed from the stores unless they had purchasing tickets rationed by the state. This system, which covered about forty basic commodities, was abandoned in 1986. Moreover, the state no longer regulated prices for the sale of agricultural products so that the proportion of farm products sold at state-regulated prices dropped from an overwhelming 91.3 percent in 1978 to only 23.9 percent in 1986. Within the same period, the proportion sold at fluctuating market prices increased from 8.7 percent to 76.1 percent. Similarly, the retailing of daily used goods, formerly determined by state-regulated prices, accounted for 97.2 percent of total sales in 1978. After pricing was aligned with the free market, the proportion of free market sales increased from a mere 2.8 percent in 1978 to 86.5 percent in 1986 (Zeng 1988:2-3).

With all of these changes, the state is no longer responsible for the purchase, supply, and distribution of agricultural products and manufactured goods. Agricultural production is now regulated by market forces and farmers are able to respond to the demands of the market according to their own production capability. This change has enabled farmers to specialize in crops that they perceive as profitable and for which they have a comparative advantage. It has also required farmers to be flexible in response to the changing demands of the market.

Finally, the reformed system of taxation and local finance has given Nanhai greater local autonomy and motivation to mobilize capital from all possible sources for the development of the township and village enterprises. Under the old financial system of 'state monopolizing revenue and expense' (*tongshou tongzhi*), almost all revenue generated at a local level had to be passed to higher-level governments, which would in turn allocate funding for major development projects at local levels. This old system has been abandoned and replaced in Guangdong by the new 'Financial Responsibility System' (*caizhen baogong*), which sets a fixed multi-year lump-sum revenue quota to be passed on to the higher-level government by all local governments. Any revenue generated above and beyond the quota could be retained for local expenses (Zeng 1988:2). This new system has given Nanhai greater incentives for revenue generation and freed it from the tight financial control of both the municipal government of Foshan and the provincial government of Guangdong. The implementation of this new fiscal policy has necessitated and enabled the development of manufacturing and business, because they normally generate higher revenue and more job opportunities than agriculture.

Agricultural Development

The new policies outlined here are essentially an indication of the tacit laissez-faire attitude of the reformed central state in dealing with local economic affairs. Implementation of these policies has allowed free market forces to shape the local economy. One direct and immediate outcome of the reforms has been the rapid diversification and commercialization of agriculture.

After farmers are allowed to make their own production decisions and maximize their profits, they quickly take the initiative in responding to the demands of the market. Since the demands of the agricultural market cover a wide range of products extending beyond grain, and since the cultivation of rice is generally less profitable than raising fish or planting fruit trees, the first move that farmers made was to shift the focus of farming away from rice cultivation to the production of more profitable market-oriented commercial crops. Many rice fields were transformed into fish ponds or orchards. Between 1980 and 1990, the acreage devoted to rice was substantially reduced by 42,933 mu (7,071 acres). At the same time, the fish-pond acreage increased by 10,205 mu (1,680 acres), and the orchard acreage expanded by 20,263 mu (3,337 acres) (Nanhai, Statistical Bureau 1991:15). Other commercial farming activities, such as breeding poultry, rearing pigs, and growing vegetables and flowers for urban consumption, also expanded rapidly. By 1990, traditional grain production no longer dominated agriculture. Its share of the total value dropped from 53 percent in 1980 to only 34 percent in 1990 (Nanhai, Statistical Bureau 1991:9).

The restructuring of agricultural production has two remarkable features. First, it is shaped primarily by free market forces. This fact is evident from the allocation of farmland, which is now frequently changed according to the fluctuations of the market. When the market price for fish was first deregulated in 1981, for example, the area devoted to fish ponds increased substantially by 2,320 mu (382 acres) in that year (Nanhai, Statistical Bureau 1989:10). Similarly, when the demand for fruit shot up in the mid-1980s, orchards suddenly expanded by 23,370 mu (3,849 acres) in the two years from 1985 to 1987 (Nanhai, Statistical Bureau 1989:10). Generally, after a particular crop is perceived as more profitable, farmers immediately decide to expand production of that crop until the economic returns diminish as the market becomes saturated. Thus, the cropping pattern and agricultural activity type being carried out, which formerly were determined by the state, are now under the control of the market.

Second, the process of agricultural diversification and commercialization in Nanhai was greatly facilitated by its strong connections with nearby large cities and the possibility of exporting. Since the reforms, Nanhai has built a number of export production bases that specialize in the production of fruit, vegetables, pond fish, and fowl. By 1987, ten such production bases had been built, covering 340,000 mu (55,998 acres), and accounting for 40 percent of the total farmland. They had generated an export revenue of US$18.98 million, about 70 percent of Nanhai's total agricultural export (Nanhai, Work Team 1988a:8-9). Some of these export bases operated over a fairly large area. For instance, the poultry farm in Shishan zhen (township) covered 1,544 mu (254 acres) (Nanhai, Work Team 1988a:11). The vegetable farm in Lishui zhen (township) was even bigger, covering 5,000 mu (823 acres) and annually producing 20,000 tons of high-quality vegetables (Nanhai, Work Team 1988b:14-15). These huge farms were equipped with modern agricultural machinery, most of it imported. Taking full advantage of the economies of scale, these farms have generated a net income higher than that of individual smaller farms. Such a new agricultural production system, called by the local people 'management on a sizeable scale' (*shidu guimo jingying*), is becoming popular in the countryside and other places in the delta region. This pattern of agricultural development, where large-scale modern farming is gradually taking shape, stands in sharp contrast to the pattern of persistent small-holding agriculture in the lower Yangtze Delta (Huang 1990:18). It appears that the existence of nearby large urban centres has enabled agriculture to be diversified, commercialized, and modernized more rapidly in Nanhai than elsewhere.

While many rice fields are being turned over to the production of market-oriented commercial agriculture, farmers cannot completely stop rice production, because they still have to meet the contracted minimum rice quota to keep the land assigned to them under the Household Production Responsibility System. Farmers in Nanhai, however, have found ways to meet their grain quotas while engaging in other more profitable activities. For example, some farmers whom I interviewed in Dali zhen have negotiated with local officials and since 1984 have been allowed to make cash payments in lieu of the rice quota. Other farmers have begun to hire people from the poorer areas outside the delta region to cultivate their rice crops, thus meeting their quotas, while they themselves have taken higher-paying jobs off the farm.

Another increasingly popular solution has been to let those farmers who are not good at rice cultivation subcontract their allocated land to farmers who are 'rice-growing specialists' (*zhong tian nen shou*). In some localities, such as the Dong Village of Pingzhou zhen (township), all rice fields with an area of 145 mu (24 acres) were contracted to the household of Farmer Zhu Jingcheng, a rice-cultivation specialist. Zhu and his family used their expertise to the greatest extent and made a higher profit from rice cultivation than they otherwise could have made. The grain produced by the Zhu's household was then sold to the local villagers for their own consumption and for meeting their required grain quotas. By so doing, other village farmers were freed from rice production and were able to do whatever they felt was more profitable and suitable to their own expertise. In this way, agricultural production has become increasingly specialized.

The process of agricultural specialization has gone beyond the sphere of production and has begun to affect the marketing and circulation of farm products. Under the Household Production Responsibility System, farmers could sell their surplus products after the contracted quotas were met. During the harvest season, many farmers found it more efficient to sell the farm products immediately to an intermediary who would resell them at retail prices, or simply hire someone with a vehicle to transport their goods to market. New trade and transport specialists have, therefore, emerged to take advantage of these new opportunities. Some individuals have bought a tractor or motorcycle and specialized in transporting goods. A group of motorcyclists or tractor drivers often waits at the entrance of a rural village or town for possible transport jobs.

The diversification and specialization of agricultural production is also manifest in the variety of occupations for the peasant family. In the pre-reform era, all adult workers of a family had to work for the production team to get sufficient 'work points' (*gong fen*) for its grain allocation. Part of the involutionary growth process of the era, farmers had no other options, regardless of their talent or inclination. Under the Household Production Responsibility System, two persons in the family, often one of the elderly and the housewife, can easily handle all the farm work on the field the household is allocated. Other family members, the husband and his sons or daughters, are free to take more lucrative jobs elsewhere. Consequently, it is not uncommon to find the members of a peasant family having a variety of occupations, including farming, manufacturing, transport, construction, and trading.

Along with the diversification and specialization of agriculture, the traditional cultivation of rice has begun to enter a new stage of mechanization.

In the pre-reform era, the mechanization of agriculture made little progress, despite the rhetorical campaign of the central state. Ironically, it was after the de-collectivization of the agricultural production system that the cultivation of rice became spontaneously mechanized. This process of agricultural mechanization can be illustrated by development in the rural district of Liangjiao, where I did fieldwork in 1992.

Located in the midst of a flat plain between the cities of Foshan and Guangzhou, Liangjiao qu (district) is traditionally an area for paddy rice production. It consisted of two villages and had a total population of 2,456 in 1992. The implementation of the Household Production Responsibility System in the early 1980s divided Liangjiao's 1,300 mu of cultivated land into small tracts for individual peasant households. While the de-collectivization of farming did provide incentives for farmers to work harder for higher personal gains, it made it difficult for efficient use of the irrigation system and other agricultural machinery. In 1991, the leaders of Liangjiao qu decided to do a little experiment. From the existing cultivated land of 1,336 mu, they took 300 mu to form a large farm for rice cultivation. The farm was looked after by twelve experienced farmers who were good at rice production. A well-regulated irrigation system was built. High-quality grain seeds were sown. Modern farming machinery imported from Germany was used. The experiment turned out to be a great success, with a grain yield exceeding 500 kilograms per mu. The experimental farm was then expanded into a huge one covering 1,000 mu (164.7 acres). For better management, an agricultural development company was formed in 1992. The company hired twenty farmers who worked eight hours per day, six days per week, for a salary of 500 yuan per month. A bank loan was made available to the company by the Agricultural Bank of China for purchasing machinery, grain seed, and fertilizer, and for other farm infrastructure investment. The company signed a contract with the district of Liangjiao that required it to meet an annual grain production quota of 850 kilograms per mu (double-cropping). Anything produced above the quota could be retained by the company.

As a result of mechanized and intensive farming, the grain yield far exceeded the contracted quota by over 100 kilograms per mu in the 1992 spring crop, which resulted in a net income of 146,080 yuan for the company in the first half of the year. With double-cropping, the company was able to make a net annual income of 290,000 yuan (US$60,669).

To what extent and in what manner has the process of agricultural restructuring contributed to the process of spatial transformation? My field investigation in Nanhai suggests that agricultural restructuring as part of the process of change does not act alone in creating major spatial patterns. Instead, it is the interaction of agricultural restructuring with rural industrial development that has shaped the spatial pattern of change in population redistribution and land-use transformation. More specifically, agricultural restructuring has interacted with industrial development at the township and village levels in the following ways.

First, the diversification and commercialization of agriculture has significantly increased rural income and provided substantial capital to facilitate rural industrial development. The financial advantage of cash-crop farming over rice cultivation has been remarkable. According to a survey conducted by the Agricultural Commission of Nanhai in 1986, cultivating rice can generate a net income of only 195 yuan per mu, much lower than the net income generated by raising fish (546 yuan per mu) and planting sugar cane (303 yuan per mu), not to mention growing vegetables, fruit, or flowers which normally produces a net income as high as several thousand yuan for the same area (Nanhai, Agricultural Commission 1987:10). The shift in production focus from traditional cheap grain to more profitable cash crops has, therefore, enabled the value of agricultural production to rise substantially. Rural per capita income rose sharply from 350 yuan in 1980 to 1,701 yuan in 1990. The income generated by peasants in 1990 on a per capita basis was significantly higher than either the regional average of the delta (1,288 yuan per person) or the provincial average of Guangdong (1,043 yuan per person). It was even higher than that of other economically advanced counties such as Shunde (1,500 yuan per person), Dongguan (1,359 yuan per person), and Zhongshan (1,531 yuan per person).

With a growing peasant income, personal savings shot up from a total of 52.12 million yuan in 1978 to 1.658 billion yuan in 1987. Per capita savings deposits rose dramatically from a 136 yuan in 1980 to 4,359 yuan in 1990, again much higher than the regional average of the delta (2,713 yuan per person) and the provincial average of Guangdong (1,205 yuan per person). Such sizeable bank savings in combination with overseas remittances estimated at US$69.72 million from 1978 to 1987, have provided a great amount of capital for investment in the industrial sector. In 1986, for instance, a total of 900 million yuan was provided through bank loans for the development of industries in the townships and villages of Nanhai (Byrd and Lin 1990:78).

Second, the specialization and mechanization of agricultural production has released a great number of surplus rural labourers who have turned to rural industry for employment. The intrusion of market forces has greatly motivated peasants to increase labour productivity so as to maximize personal gain. Data from Nanhai have shown that labour productivity has, indeed, significantly improved since the reforms, with the agricultural value per labourer from 1980 to 1990 rising from 1,005 yuan to 4,650 yuan at the 1980 constant price (Nanhai, Statistical Bureau 1989:5; 1991:10). In the rural district of Liangjiao where I conducted fieldwork for this book, before mechanization the farming of the available 1,336 mu required the entire labour force of 1,316 persons, but after mechanization, the cultivation of rice on 1,000 mu required only twenty experienced farmers. When these farmers and other administrative staff as well as emigrants were deducted, a total of 1,184 peasants, over 90 percent of the total labour force, was no longer needed on the farm (Dalizhen 1992c:5). For Nanhai as a whole, it was reported that by 1987, as many as 237,600 peasants, representing 65 percent of its total rural labour force, had been removed from agricultural production (Zhao 1988:355). These unemployed peasants had to find non-farm employment in the townships or villages of the countryside. Thus, the mechanization and specialization of agriculture has generated a large number of surplus rural labourers as potential labour for the development of rural industries.

Third, the commercialization and mechanization of agriculture has raised peasant demand for more manufactured goods, including not only agricultural equipment, machinery, and other supplies for production purposes, but also many consumer products, such as motorcycles, televisions, electric fans, washing machines, and refrigerators. The influence of Hong Kong through radio and television has further stimulated a desire for modern consumer goods among the peasants. State factories in large cities were encouraged to meet this new demand, but after three decades of production following the Soviet model which overemphasized heavy industry, it is difficult for the big old plants in large cities to convert their machinery and retrain workers for the production of new consumer goods. By comparison, numerous small-scale rural industries can respond with greater speed and flexibility in spite of their unsophisticated technology. Thus, the increasing demand of peasants for new consumer goods has directly fuelled the rapid growth of consumer industries in the countryside.

Finally, the diversification and specialization of agriculture has revitalized the commercial function of small towns, which are sparsely distributed over the countryside. As the surplus of market-oriented farm goods has increased, rural fairs and markets have not only been re-opened but also held more frequently than ever before to meet the demand of local peasants for buying and selling. The total value of Nanhai market sales jumped from 326.25 million yuan in 1980 to 16.45 billion yuan in 1990 (Guangdong, Statistical Bureau 1991b:173). Many markets are constantly expanding and some have begun to specialize in trading certain farm goods or commodities. For example, the poultry market in Dali zhen was one of the first and largest specialized markets in Guangdong Province. Its daily sales volume has reached 4,000 fowl and 13,000 kilograms of vegetables (Zhao 1991:169). The revitalization of the commercial function of small towns as a result of agricultural commercialization has become a significant impetus promoting the development of these towns.

It may be argued from the foregoing analysis that the diversification, specialization, and commercialization of agriculture, which occurred at the grassroots of the rural economy, has significantly contributed to the process of rural industrialization, although the restructuring of agriculture per se did not directly result in major spatial development. The two processes of agricultural restructuring and of rural industrialization have been so closely intertwined that it would be highly inappropriate to isolate one from the other. In this regard, the experience of Nanhai is distinct from that of other parts of the nation where market farming is unable to boost rural industrial development because of limited access to overseas international markets (Huang 1990). Nanhai, however, is similar to other parts of the country in that industrial development at the township and village level has been the most powerful force directly shaping the spatial pattern of population redistribution, land use, and town growth in the post-reform era.

Rural Industrialization

The rapid growth of rural industry at the township and village level since economic reform has been a national phenomenon, but rural industrialization in Nanhai is distinguished by its magnitude and approach. As described by the local people, Nanhai is renowned for its distinct approach of 'driving forward on five wheels,' which encourages simultaneous industrial growth at all levels, including county, township, district, village, and private partnership. Before the pattern of rural industrialization is assessed, it is necessary to clarify the local administrative structure, as it has significant implications for the growth and distribution of rural industries.

The administrative system in Nanhai consists of five basic components ranked in descending order from county (*xian*), township (*zhen*), administrative district (*guanli qu*), village (*cun*), to village community (*cunmin weiyuanhui*). Needless to say, the county is the top of the hierarchy, with the highest authority for managing local economic affairs, including personnel appointments, tax collection, and budget allocation. Next, a number of townships have recently been created to replace the rural communes that existed in China from 1958 to 1984. On average, a township has a population of 55,000 and an area of 68 square kilometres. A typical township has a town seat, which is normally accorded a designated town status, although the overall population and land area of the township are predominantly agricultural in nature. Below townships are several administrative districts formed in 1984 to replace the brigades. A typical district has a population of several thousand and usually consists of a couple of villages that were formerly production teams. At the bottom of the hierarchy are the village communities, which are mainly for census and organizational purposes. These communities have lost their economic functions as a result of the implementation of the Household Production Responsibility System.

In all, Nanhai had 17 townships, 242 administrative districts, and 1,406 village communities in 1990. Under this administrative framework, rural industries or township and village industries (*xiangzhen gongye*) are actually the descendants of former commune and brigade industries (*shidui gongye*). The term 'rural industries,' however, is used in this study because it includes not only those industries owned by townships and villages but also those owned by private partnerships or individual households, which have become increasingly important in the delta region, especially in Nanhai.

Small-scale rural industry, as a means to aid agricultural production, had existed in Nanhai before the economic reforms were introduced in 1978, but such industries had never played a major role in the local economy. Under the then prevailing policy, which overemphasized grain production at the cost of other non-farm activities, the county government had even limited the growth of rural industries by ruling that commune and brigade enterprises should not hire more than 10 percent to 15 percent of the local labour force. It was not until the early 1980s that rural industries were allowed to flourish. Incidentally, in 1980, the county government noted that in one commune where industrial enterprises were tolerated and allowed to develop freely, peasant income had become higher and increased more rapidly than elsewhere. County officials reacted positively, encouraging commune and brigade enterprises to develop across the county. Local people quickly seized the opportunity, and one year later 71 percent of Nanhai's production teams started their own industrial enterprises (Byrd and Lin 1990:152). These industrial enterprises remained collectively owned until 1983 when the production responsibility system was introduced to de-collectivize production. Many production team enterprises were leased or sold to individuals for private management. In that one year, the aggregate value of the fixed assets of production teams dropped from 158 million yuan to 134 million yuan, while the value of privately owned fixed assets rose from 16 million to 104 million yuan (Byrd and Lin 1990:152).

The privatization of industrial enterprises linked investment and labour input directly with personal gain and, thus, greatly motivated peasants to run factories efficiently and successfully. As a result, rural industries expanded rapidly to become a major pillar of the local economy. In March 1984, the commune-brigade-team system was officially dismantled and the 'commune and brigade industry' (*shidui gongye*) was renamed the 'township and village industry' (*xiangzhen gongye*).

The growth of rural industry since the reforms and especially since 1984 has been spectacular. Nominal gross income generated by rural industry increased thirteenfold from 342.4 million yuan in 1980 to 4.67 billion in 1990 (Nanhai, Statistical Bureau 1991:13). The share of rural industry in the total industrial production of the county rose from less than a quarter in 1978 to 66.56 percent in 1991 (Nanhai, Bureau of Township and Village Industries 1992:3). In 1978, peasants who were engaged in industrial and sideline production accounted for 25.4 percent of the total labour force. By 1991, a total of 246,153 jobs had been generated by the rural industrial sector alone, accounting for 60.88 percent of the total labour force. The tax revenue generated by the collectively and privately owned enterprises, most of them industrial, accounted for 70 percent of the total 1988 tax revenue of the county, far exceeding that of the state sector, which was only 27 percent (Nanhai, Statistical Bureau 1989:72). Obviously, rural industry has become the backbone of the local economy and has played a decisive role both in meeting the rising demand for consumer goods and in competing with state-owned industry.

Rural industries developed in Nanhai are predominantly labour intensive and market oriented. By far the largest group of factories deal with textiles and apparel, in which Nanhai has traditional strength. Since the silkworm cocoon used to be a major agricultural product, it is not surprising to find that silk processing and, later, the textile industry have become its chief industry. The first modern silk mill in China was built in Nanhai's Xiqiao zhen (township) at the turn of this century (Zhao 1991:45). By 1992, more than 1,000 textile factories operated in Nanhai, with fixed capital assets of one billion yuan and over 20,000 textile machines (Yang 1992:8). These factories use a variety of materials, including polyester fibres, cotton, and silk, and cover all phases of the production process. Their annual production capacity has exceeded 200 million metres of cloth. The next largest group of factories focuses on metal manufacturing, particularly aluminum processing. With technical assistance from Germany, Italy, and Japan, 82 factories for aluminum processing have been built since 1986, including a modern factory, the only one of this type in southern China, producing thin aluminum metal (Yang 1992:9). The third largest group concentrates on ceramic tiles, for which Nanhai has plenty of material resources and an excellent tradition of production. The expansion of ceramic tile production has coincided with the booming of real estate in Guangdong and other parts of China, opening a large market for tile production. By 1992, Nanhai had 59 tile factories, over 100 ceramic production lines, with a nationwide sales force and an annual total production capacity of 50 million square metres of tile (Yang 1992:9). In addition to the above three sectors, numerous small factories produce all sorts of household consumer goods such as toys, shoes, watches, cans, wines, electric fans, and microwaves. Most of the production is to satisfy the growing demand of domestic markets, but some products, particularly toys and shoes, are manufactured based on capital investment and technology from Hong Kong and are mainly for export.

Several features are associated with the development of rural industries. First, many factories in the townships and villages are fairly small. Data obtained from local authorities indicated that 10,865 industrial enterprises operated at the township and village level in 1991. These industrial enterprises varied in size and type, but, on average, a factory employed only twenty-two workers, which is fairly small by Chinese standards (Nanhai, Bureau of Township and Village Enterprises 1992:5). This small size makes economies of scale difficult to achieve. It has, however, allowed for great sensitivity and adaptability to changing market demands. Geographically, these small-sized factories can be located almost anywhere, as they do not require massive infrastructure investment. Thus, it is not uncommon to find small factories dotted across the countryside, sometimes in the middle of rice fields.

Second, while rural industries have flourished at all levels of the rural-urban hierarchy since the reforms, they have emerged primarily from the grassroots. When the production of rural industries is broken down according to the four major ownership forms classified by local authorities, industrial enterprise at the village level has stood out as the leading player in production and in employment generation (see Table 6.1). Enterprises developed by townships, individuals, and private partnerships have done their share, but lag behind the contribution of village enterprises. When other enterprises, whether state-owned, foreign-owned, or joint ventures, are included for comparison, the contribution of village industrial enterprises to the total industrial production remains outstanding. They produced 40 percent of the total industrial output value in 1991, far more than the 14 percent share of the state sector, the 38 percent of the township enterprises, or the 6 percent of the foreign firms or joint ventures (Nanhai, Statistical Bureau 1992:44). Such a pattern of production

Table 6.1

Industrial ownership structure for Nanhai, 1991

Ownership type	Output value (%)	Income (%)	Employment (%)
Township	39.02	38.60	28.04
Village	44.73	44.95	46.67
Partnership	4.30	4.36	7.64
Individual	11.95	12.09	17.65
Total	100.00	100.00	100.00

Source: Nanhai, Bureau of Township and Village Enterprises 1992:1-52.

suggests that industrial development is primarily shaped by the enterprise of rural villages. It also suggests, from the production point of view at least, that the chief player of rural industrialization is neither the state sector nor the privately owned individual enterprises, but the 'collectively owned' village industries. It should be stressed, however, that the 'collective sector' is not the same as that established in the Maoist era. Many 'collectively owned' industrial enterprises have actually taken the form of corporations and are run by local cadres who form a board of directors or even hold shares of the industrial assets.

Third, the growth of rural industry is essentially spontaneous and self-sustained, driven primarily by local initiative. Capital investment in the rural industrial sector has been largely mobilized from local resources through a number of channels. The local branches of the Agricultural Bank of China hold the bulk of household savings deposits of the county. It has provided a considerable number of bank loans to rural communities for setting up industries. From 1980 to 1986, the Bank of China lent an estimated US$7.8 million, mostly short term, to rural industrial enterprises in Nanhai. In 1986, these industries obtained total bank loans of 900 million yuan through various channels (Byrd and Lin 1990:78). Remittance from friends and relatives in Hong Kong and overseas also contributed to capital investment in the rural industrial sector. By contrast, the state has not been actively or directly involved in financing rural industrial development in spite of its rhetoric of encouragement. In fact, the state contributed only 16 percent to the total investment in fixed assets in 1991. The remaining 84 percent was realized by collective and private sectors through various local initiatives (Nanhai, Statistical Bureau 1992:76). With regard to raw materials, local people have reported that they no longer depend on state planning for needed raw materials and that mandatory state plans cover only one-eighth of their total industrial output. Even the technology and production methods have been obtained by local people through their contacts with firms in Hong Kong and overseas. Thus, except for being a property owner who maintains the power of tax collection and personnel appointment, the state has contributed little to the provision of capital, technology, or raw materials for rural industrial development. It is local initiative, mainly at the township and village level, that has driven the process of rural industrialization.

Spatial Transformation
The rapid growth of rural industries has significantly facilitated the process of structural and spatial transformation. The most remarkable effect of industrial development in the countryside has been the creation of employment to absorb surplus rural labour released from agricultural production as a result of increased labour productivity. It was reported that

from 1985 to 1989 an average of 13,534 jobs were created each year by rural industries to accommodate surplus rural labourers (Nanhai, Bureau of Township and Village Enterprises 1992:3). Whereas 40 percent of the agricultural labour force in 1978 was unable to find jobs outside agriculture, by 1986 Nanhai had, for the first time in its history, a labour shortage. Outside workers had to be recruited to control the rising labour costs resulting from this shortage. Consequently, Nanhai was changed from a county of out-migration, because of underemployment in the pre-reform period, to a destination favoured by immigrants. When the first national population census was conducted in 1982, Nanhai recorded a net population loss of 1,382 persons. This pattern reversed in 1990 when immigrants outnumbered emigrants by 99,893 (Nanhai, Population Census Bureau 1992a:22). Thus, the local economy has been fundamentally transformed from underemployment, or 'involutionary growth,' to increased wealth, improved labour productivity, and abundant employment opportunities, or 'transformative development' (Huang 1990:13, 18). The driving force underlying this process is clearly rural industrialization.

More importantly, the development of rural industry has enabled a growing number of peasants to 'enter the factory without moving into the city' (*jinchang bujincheng*), or 'leave the soil but not the village' (*litu bulixiang*). According to population census data obtained from Nanhai, the occupational structure of the local population in 1982 was predominantly agricultural. A great majority (58 percent) of the population was engaged in agricultural production and the number of factory workers accounted for no more than 29 percent. Eight years later in 1990, the situation had been fundamentally changed. The proportion of factory workers in the total population jumped from 29 percent to 41 percent, whereas the agricultural share dropped from 58 percent to 38 percent (Nanhai, Population Census Bureau 1992a:52). Obviously, a substantial increase in factory workers took place at the expense of agricultural labourers. In other words, a considerable number of farmers left the soil and entered the factory. Most of the peasants who entered factories continued to live in the countryside, as is evident when the number and proportion of factory workers are compared with those of town residents. According to the 1990 population census, while there were 248,672 factory workers in Nanhai, 41 percent of its total population, there were only 172,557 town residents, 16 percent (Nanhai, Population Census Bureau 1992a:26, 52). The number of factory workers far exceeded the total number of town residents. The excessive number of workers must be those who worked in the factories outside of the towns. As town residents were not entirely engaged in manufacturing, the actual difference between the total number of factory workers and those workers who resided in the towns must be substantial. Thus, a considerable number of factory workers worked and lived in the countryside. This finding is consistent with that of the previous analysis, which revealed that township enterprises accounted for only a small portion of the industry and that the focus of industrial development was in the rural villages. It also suggests that the success of industrial development in the countryside has been able to deter a massive rural exodus into the city. Rural industrialization appears to be a crucial factor that explains why there is no significant population concentration in the large cities of the Pearl River Delta, despite rapid industrial development as identified in Chapter 5.

The growth of rural industry has also contributed to changes in land use. As manufacturing is perceived as a desirable activity that can generate more jobs and higher income than rice cultivation, a considerable amount of farmland has been transformed into paved industrial sites. It was reported that from 1982 to 1990, a total of 3,597 acres of farmland was taken by non-agricultural development. Of this lost farmland, industrial expansion accounted for about 32 percent, transport development about 21 percent, residential land use 19 percent, and other construction purposes the remaining 26 percent (Chen 1992:5). Industrial development stood out as the leading sector responsible for most, albeit not all, of the loss in agricultural land.

It is not possible to show exactly how much industrial land expansion was in the countryside and how much was in the towns. It is known, however, that most factories have been built in rural villages rather than in existing towns. Of the 11,109 industrial enterprises that have been built, 10,524 or 94 percent are found in or below the village level (Nanhai, Statistical Bureau 1992:44). Most of these factories are located at the entrance of a village or along the highway that passes by the village. Such locations allow for the easy transport of both raw materials and finished products. Some factories can be found on the edge of the village immediately next to rice fields. By building a factory within the jurisdiction of the village, factory owners (that is, the peasants of the village) are able to save a considerable amount of land rent, as rents are much higher for city or town sites. As well, such a location is easily accessible for local villagers who have entered the factory but not the city.

As a result of rapid industrial growth at the village level, there is growing mixed land use by the industrial and agricultural sectors in the countryside. Rural industrialization has, thus, gradually created a new landscape in which industrial and agricultural, or urban and rural, land uses stand side by side, leading to a blurring of the rural/urban distinction. The experience of Nanhai suggests that rural industrialization is one of the most important factors underscoring the pattern of land-use change identified in Chapter 5, where the transformation of land use in the Pearl River Delta region was found to be taking place more in the countryside than in the cities.

In addition to changes in population distribution and land use, rural industrialization has created some significant environmental consequences, which have usually been neglected by planners who are overwhelmed by the environmental problems in large cities. As shown, industrial production is mostly on a small scale with unsophisticated technology for simple manufacturing. Many of these factories do not have the necessary facilities and advanced techniques for the proper treatment or recycling of industrial waste, because the factories are simply too small to afford such modern facilities. The location of the factories, which is virtually in the 'grey area' of the countryside where environmental regulations are looser than in the city, has further enabled unregulated and untreated disposal of industrial waste. Moreover, as hazardous and polluting industries are no longer tolerated in large cities, Nanhai, a suburban county immediately next to the cities of Foshan and Guangzhou, has increasingly become a major target for relocation by heavily polluting industries including sugar refining, cement manufacturing, textile printing and dying, electroplating, and aluminum processing. Consequently, much of the Nanhai area has been treated as a dumping ground for various waste materials generated by industrial production.

An early survey conducted by the local environmental agency revealed that industrial production in 1988 released a total of 4,166.3 tonnes of sulphur dioxide, nitrous oxide, carbon dioxide, and particulates into the air (Nanhai, Environment Monitoring Station 1989:69). The emission of these hazardous materials skyrocketed in 1990 to 60,737.56 tonnes as a result of the rapid growth of rural industry (Nanhai, Environment Monitoring Station 1991:33). At the same time, the amount of waste water discharged from industrial production jumped from 52.7085 million tonnes in 1988 to 135.637 million tonnes in 1990 (Nanhai, Environment Monitoring Station 1989:70; 1991:34). Most of the waste-gas emission came from the combustion of poor-quality coal, which provided almost all the energy and electric power for rural industry. The waste water was generated primarily by factories involved in sugar refining, pulp and paper processing, textile dying, and electroplating. Such waste water was often released directly into streams that provided water for rice or vegetable fields, and even into fish ponds, without proper treatment or purification, leading to the serious contamination of farmland and crops. In Lidong qu (district) of Dali zhen (township), for instance, of the ninety mu of fish ponds, seventy mu were contaminated. On average, one out of every five mu of cultivated land in Nanhai was found to be contaminated (Guangdong, Foshan, and Nanhai, Joint Investigation Team 1989:16-17).

Similarly, increasing waste-gas emission has caused severe air pollution and resulted in some damage to the local ecosystem. Among other problems, acid rain has been reported to occur more frequently than ever before, with the frequency of occurrence rising significantly from 1.6 percent in 1986 to 17.49 percent in 1990 (Nanhai, Environment Monitoring Station 1989:6; 1991:4). Some of these ecological changes caused by the devastating action of industrial development could be irreversible and disastrous.

While rural industrialization has brought with it considerable wealth for the peasants, it has, simultaneously, degraded the environmental quality of life for the local people, and reduced the sustainability of the environment for further economic expansion. Given that the local economy is currently on an upswing and that the people of Nanhai are preoccupied with seeking higher profits at any cost, they will not likely stop exploiting natural resources or make sacrifices to preserve a sustainable environment for the sake of future generations. On the contrary, as industrial and agricultural production continue to expand, natural resources and the environment will likely suffer even more to satisfy the ever-growing demand of the local people for wealth and consumer goods. Therefore, rural industrialization, which is essentially unplanned, small scale, and intimately related to the natural environment, has been and will continue to be one of the most powerful forces changing not only the spatial economy of the region but also the natural ecosystem of the human habitat.

Summary

The Chinese countryside has experienced profound change since economic reforms were initiated in 1978. While it is generally known that the reforms have brought much wealth and prosperity for the Chinese peasants, the operating mechanism of this process and its subsequent spatial and environmental consequences remain little understood. This case study of Nanhai reveals that the transformation of the rural economy in the Pearl River Delta is characterized by two simultaneous processes: agricultural restructuring (that is, commercialization, specialization, and mechanization) and rural industrialization. These two processes are found to be so closely intertwined that it would be highly inappropriate to isolate one from the other.

Although the penetration of free market forces has led many profitable non-agricultural activities to assume an increasingly important position in the rural economy, agriculture as a traditional economic sector has not been completely wiped out by the newly emerging, modern, urban-based industries. Instead, the experience of Nanhai has demonstrated that agriculture could be restructured to meet the growing and diversified demands of the market. Consequently, agriculture not only has continued to exist but also has become so prosperous that a considerable financial surplus has accumulated which, in turn, has boosted rural industrial development. This process has led to the formation of what some local Chinese scholars call 'a double-dualist structure' in which agriculture and industry, urban and rural, sectors stand side by side (Zhao 1991:185).

Unlike the experience of countries in Southeast Asia where the persistence of agriculture as an economic sector for survival is due to the fact that industrial expansion was unable to keep pace with population growth, the persistence and recent development of agriculture in Nanhai are primarily shaped by other factors. These factors include a well-established farming tradition, excellent natural endowments, and, most importantly, easy access to the large urban markets in Hong Kong, Guangzhou, and Foshan, which have kept market farming a profitable activity for the local people. In this regard, the experience of Nanhai has been significantly different from that of other parts of China, such as the lower Yangtze Delta, where market farming has contributed little to the improvement of the peasant economy, possibly because of limited access to overseas international markets (Huang 1990).

While the commercialization of farming has contributed to higher household income and personal savings for peasants, rapid industrial development in the countryside is found to be the most powerful force directly responsible for many of the structural, spatial, and environmental changes that have occurred. By creating a great number of factory jobs, rural industries have absorbed a substantial amount of the surplus farm labour released from agricultural production. In this manner, rural industrialization has fundamentally transformed the local economy of Nanhai from one of prolonged 'involutionary growth,' or 'growth without development,' into one of unprecedented genuine development. Rapid industrial development in the countryside has also enabled many local peasants to 'enter the factory but not the city,' thus preventing a massive rural exodus and corresponding urban influx. Geographically, the flourishing of numerous small-scale industries in the countryside has facilitated the rapid encroachment of industrial development over valuable farmland, leading to a mix of intensive land uses including agricultural and industrial production, or urban and rural activities. The primitive nature of such industrial development, and its 'hidden' location in the countryside, has favoured an unfriendly treatment of the natural environment, which has caused serious damage to the local ecosystem.

The recent phenomenal growth of production in both market farming and rural industry has been primarily a result of the state's relaxed control over the local economy, not a consequence of any active state intervention. Statistical data from Nanhai have clearly indicated that the state sector or publicly owned enterprise accounted for only 16 percent of total capital investment, 13 percent of industrial output value, 28 percent of employment, and 27 percent of the total tax revenue (see Table 6.2). The private sector was not the chief economic player either, as its contributions to local

Table 6.2

Sectoral composition for the local economy of Nanhai, 1991 (%)

Items	State sector	Collective sector	Private sector	Total
Capital investment[1]	16.62	63.98	19.40	100.00
Employment[2]	28.39	62.74	8.87	100.00
Industrial output value[3]	14.22	71.42	14.36	100.00
Tax revenue[4]	27.46	60.11	12.43	100.00

Sources:
[1] Nanhai, Statistical Bureau 1992:76.
[2] Nanhai, Statistical Bureau 1992:100.
[3] Nanhai, Statistical Bureau 1992:6, 54. Raw data are in 1990 constant prices. The private sector includes joint ventures whereas the collective sector includes all collectively owned enterprises in towns and villages as well as those in partnerships.
[4] Nanhai, Statistical Bureau 1989:72. Data are for 1988.

financing, industrial production, employment, and tax revenue have been smaller than those of the state sector or the collective units. By comparison, township and village enterprises, which are statistically classified as 'collectively owned' but are actually operated by local cadres who form corporations, have increasingly become the mainstay of the local economy. They have contributed 64 percent of total investment, 63 percent of employment, 71 percent of industrial output value, and 60 percent of total tax revenue (see Table 6.2).

In view of this unique pattern, it may be argued that a new local economy, motivated primarily by local community initiative and shaped by free market forces, is gradually taking shape in the Pearl River Delta. Consequently, many features of structural and spatial transformation of the delta region identified in Chapter 5 can be attributed more to spontaneous development at the local level than to the intervention or active participation of the central state.

7
Transport Development

In the study of the growth dynamics of regional development in the Pearl River Delta since the 1980s, much scholarly attention has been directed to the process of economic reforms in the countryside, and to the consequences of implementing the open door policy. By contrast, the role of transport development in the process of spatial transformation has been overlooked, probably because large-scale construction of the transport infrastructure is a fairly recent phenomenon and detailed data are not yet available for systematic assessment. Transport development, however, is a critical factor underlying many of the spatial changes that have occurred in the delta region.

From a theoretical standpoint, despite the scholarly debate about the nature of the relationship between transport investment and economic development, geographers and planners generally believe that transportation is a key element of spatial change, which, by reducing the 'friction of distance' or 'collapsing time and space,' has direct effects on the process of spatial reorganization, in particular the concentration and specialization of economic activities and the suburbanization of human settlements (Janelle 1969:348-64; Abler 1975:35-56; Brunn and Leinbach 1991). More specifically, transport improvement was identified by Gottmann, McGee, and many others as a necessary condition for the metropolitan development that has taken place not only in the United States but also in many Asian countries (Gottmann 1961:632; McGee 1991b:16-17). It was also considered by others to be the key force that has shaped the growth and transformation of major development corridors in North America and the Pacific Rim (Yeung 1994; Rimmer 1995; Yeung and Lo 1996). The Pearl River Delta region has exhibited many features of metropolitan development, such as high population density, great mobility of people and goods, intensified rural-urban interaction, and the emergence of corridors along major transportation arteries. Given this fact, it is of theoretical importance to examine to what extent and in what manner transport development has facilitated the process of spatial transformation in this region.

The recent development of a transportation infrastructure in the delta region has also pointed to the importance of the transport sector as a key element of spatial change that deserves special investigation. Although no systematic data show the magnitude of transport development and its spatial consequences in the delta region, data at the provincial level do suggest that the construction of transportation facilities since the mid-1980s has become the top priority of the province's development agenda (*Yangcheng Wangbao* 1992, March 3; May 27; June 4; *Renmin Ribao* 1992, June 8; *Jingji Cangkao* 1993, March 7). Investment in transportation and telecommunication, for instance, increased from 337 million yuan in 1978 to 4.69 billion yuan in 1991, then skyrocketed to 8.8 billion yuan in 1992 (Guangdong, Statistical Bureau 1985:201; 1992a:238-46; *Renmin Ribao* 1992, June 8). Such investment represents a thirteenfold net increase from 1978 to 1991, outpacing the ninefold increase of total investment in the province. Consequently, the share of the transportation and telecommunication sector in the province's total capital investment rose from 15.9 percent in 1978 to 21.8 percent in 1991, and moved to the second rank among all thirteen economic sectors (Guangdong, Statistical Bureau 1992a:246).

Most of the infrastructure investment was directed to the Pearl River Delta region, traditionally the core of the province. As a result, the existing highway system of the delta region was extended by 3,470 kilometres in the 1980s, a total of thirty deep-dock harbours were constructed, six new airports were built, and modern telecommunication systems with direct access to major international networks were installed in most cities and counties (*Renmin Ribao* 1992, December 24). In addition, an American-style freeway connecting Hong Kong, Shenzhen, Guangzhou, and Macao, at an estimated investment of US$1.2 billion, has been completed (Chinese News Agency 1992:4). Such massive investment in transportation infrastructure is unprecedented in the history of the region, and has a major impact on the spatial patterns of population movement, land-use change, and human settlement distribution. A detailed assessment of transport development would be required to account for the features of spatial transformation identified in Chapter 5.

This chapter examines the role of transportation in the transformation of the delta's regional economy. The assessment will be based on a detailed case study of Panyu, where expansion of the transportation infrastructure in recent years has been considered by many local Chinese researchers most noteworthy among the counties of the region. The chapter begins with a contextual analysis of the existing economic and geographical conditions of Panyu. It then turns to the rapid expansion of the transportation infrastructure. This discussion is followed by an assessment of the impacts of such development on economic growth, migration, and land-use transformation.

Historical and Geographical Background

Among the cities and counties of the Pearl River Delta, Panyu is generally considered the most typical case of a region where heavy investment in the transportation infrastructure is seen as instrumental to the initiation of economic growth. The physical setting of Panyu closely resembles the general situation of the whole delta region. Instead of being an extensive piece of flat plain, Panyu is divided by many rivers and streams. The existence of a large water system is evident in the county's areal composition. Rivers and streams of various sizes occupied an estimated 35 percent of the total area of Panyu in 1980, much higher than the provincial average of 0.58 percent (Panyu, Statistical Bureau 1983:8; Guangdong, Statistical Bureau 1981:3). Such a physical environment, divided by many rivers and streams, has significant implications for the economic development of the county.

For centuries, the existence of abundant groundwater has been a favourable natural endowment for Panyu. Local people are presented with a good natural resource for irrigation and this fact has enabled them to engage in farming activities on a year-round basis with double-croppings being practised. More importantly, the streams criss-crossing Panyu have provided a cheap and easy means of transportation for marketing local products. Such accessibility to water transportation was extremely beneficial to the growth of the local economy during the ancient time when railways and highways did not yet exist and when waterways were the dominant means of interregional movement of goods and people.

Historical records show that when the Han Chinese first migrated to the delta from North and Central China, they moved in through the Si (West) River and settled down in northern Panyu as early as the Chin Dynasty, more than 2,200 years ago (Panyu, Office of Place Record 1989:1; *Panyubao* 1992, June 19; Zheng 1991:44-8). This settlement has made Panyu one of the earliest established counties in Guangdong Province. Although Panyu used to contain Guangzhou, which separated from Panyu and became an independent city in 1921, the new Panyu after separation remains one of the earliest established regions in the delta. Along with Nanhai and Shunde, Panyu has long been a part of a distinct, economically advanced area under the coined name of NanPanShun, even after Guangzhou was separated from Panyu. While easy water transportation was not the sole factor that shaped the long course of Panyu's historical development, its contribution to the early prosperity of Panyu was, nonetheless, noticeable.

However, the natural endowment in water transportation, which used to be an economic advantage, has turned out to be a serious obstacle hampering further development of the local economy since the 1970s, when quick and convenient door-to-door highway transportation became prevalent. Increasingly, the interregional movement of goods and people, especially the movement of goods between the delta and its export outlet of Hong Kong, has shifted away from the traditional waterway to modern railways, highways, and seaports, with container facilities for mass and rapid transportation. Panyu's reliance on waterways as the chief transportation means proved to be relatively uncompetitive, not only in attracting new investment from foreign and domestic sources, but also in marketing local products. Moreover, Panyu's geographical setting, which was partitioned by rivers and streams without efficient bridge connections, made it difficult for the road network to meet the standard of modern infrastructure in terms of efficiency, mobility, and flexibility. Consequently, economic growth was stagnant for years, even after new, favourable economic policies were introduced in 1978. Though Panyu used to stand side by side with Nanhai and Shunde as the most economically advanced counties in the delta, it has fallen from the leading ranks since the late 1970s and been replaced by areas such as Zhongshan and Dongguan, which have excellent overseas connections (Wang, Zhang, Zhao, Zhuo, and Liu 1991). It was probably because Panyu was unable to keep moving ahead along with its former partners of Shunde and Nanhai that the local people called the county 'the land of standstill' (*wuodidi*).

Local people became increasingly puzzled by the inability of their county to maintain its leading economic status. It did not take long for local cadres to find out that the poor transportation infrastructure was the key factor, albeit not the only one, that underscored the stagnant growth of the local economy. The local government of Panyu therefore has been determined, since the mid-1980s, to radically improve the existing road transportation infrastructure.

The commitment of the local government to transport development was also motivated by an astute recognition of the important potential of Panyu's geographical location. Panyu is situated in the geographical centre of the delta region. It borders the provincial capital of Guangzhou in the north, stretches along the main sea transportation route in the east, and bridges the west and east wings of the delta. More importantly, local officials recognized that after a good road network was established, Panyu could become the regional hub of road transportation for the entire delta. Panyu is the shortest route for interregional traffic between the hinterlands of the delta and Hong Kong, the chief economic core of the region. The newly completed Guangzhou-Hong Kong-Macao super-highway cuts through and intersects in Panyu. Thus, despite the unsatisfactory economic performance of Panyu in the 1970s and early 1980s, local officials saw great potential in the central location of the county. To exploit such geographical advantage would require much investment in building a modern road network accessible for both interregional and intraregional economic linkages.

Transport Infrastructure Development

After transport improvement was identified as the key to reinvigorate the local economy, all possible efforts were made, primarily by the local government, to achieve that development goal. The slogan of 'economic prosperity comes from the construction of roads' (*lutong caitong*) began to appear in the headlines of newspapers, at major construction sites, and at the entrances of towns and villages. A team of professional planners, architects, and civil engineers from the provincial institute of planning was invited and paid by the county government of Panyu to draw up a blueprint for development, first in 1989 and again in 1992. To minimize the bureaucratic procedure needed for decision-making and policy implementation, a special official organization named the Directive Department for the Construction and Management of Roads and Bridges was established in 1985 and located in the centre of the county capital of Shiqiao zhen. These measures were taken almost entirely by the local government and they were to help create an environment conducive to transport development. But the most effective action that had a direct impact on transport improvement was the mobilization and allocation of substantial capital to invest in the transport sector.

Data on transport investment are piecemeal and, in many cases, strictly confidential, which makes it difficult to provide a systematic assessment. Nevertheless, information obtained from various sources does suggest that the transport sector has, in recent years, become the chief recipient of capital investment. An estimated 900 million yuan (US$188 million) was invested in the transport sector from 1980 to 1991 (*Panyubao* 1992, June 19 and June 26). In 1992, when the fieldwork for this study was conducted, I was informed by local cadres that 280 million yuan (US$58 million) was directed to improving transportation infrastructure in that year alone.

When the changing composition of local budget allocation is assessed, the transport sector clearly stands out as the most significantly changed segment, rising from the bottom of the list to the second highest position, thus leading most economic sectors in receiving construction funds (see Figure 7.1). The amount of capital allocated to the transport sector increased from a mere 40,000 yuan in 1978 to 191.84 million yuan in 1991. The sector recorded an average annual increase of 91.93 percent, much higher than the growth rate of total fixed-asset investment at 43.52 percent (Guangzhou, Statistical Bureau 1989:395; 1992:173-4). Figure 7.1 shows the changing pattern of fixed-asset investment from the public sector.

It is clear that the local budget allocation before 1984 favoured the manufacturing sector, which was traditionally seen as the key to generating employment and income. By contrast, the transport sector was almost completely ignored in the allocation of construction capital. Since 1984, however, the transport sector has quickly emerged to occupy a prime position in local budget allocation. Although the manufacturing sector remains the number one priority on the list, the fact that the transport sector shot up from a neglected position to the second highest priority in budget allocation testifies to the commitment and determination of Panyu's officials to overcome the 'friction of distance.' The emphasis of investment in the transport sector has also become a special feature of Panyu that distinguishes it from other places in the delta region. Data obtained from the provincial authorities have shown that per capita transport investment was significantly higher in Panyu than in almost all other counties of the delta region, except for the two Special Economic Zones of Shenzhen and Zhuhai and the cities of Guangzhou and Jiangmen (Guangdong, Statistical Bureau 1987:205-10).

Figure 7.1

Changing composition of fixed-asset capital investment in Panyu, 1978-91

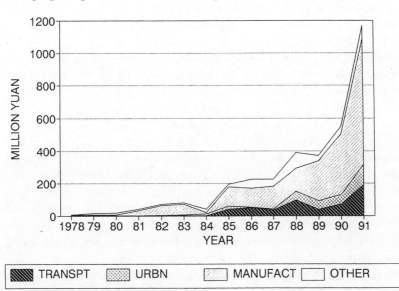

Note: TRANSPT stands for the sector of transportation and telecommunication, URBN for urban construction, and MANUFACT for manufacturing. OTHER refers to sectors including water supply and weather services, construction and survey, commerce and trade, education and health care, and other investments that are not classified in the statistics of the Chinese local expenditure.

Sources: Guangzhou, Statistical Bureau 1989:395; 1990:222-9; 1991:166-8; 1992:173-4.

Much of the capital investment in transportation was directed toward the construction of bridges and highways. It was reported that a total of 153 new bridges were built from 1980 to 1991 (*Panyubao* 1992, June 19 and 26). Simultaneously, the existing highway system was reconstructed and substantially extended, with an additional 434 kilometres of new roads added to the system, much more than the 226 kilometres of the preceding three decades (*Panyubao* 1992, June 19 and June 26). In addition, several new harbours were built, including two equipped with container facilities.

Of all the transport developments, two megaprojects have played a crucial role in creating a transactional environment. The first one involves the construction of a bridge named Ruoxi that connects Panyu with Guangzhou (see Map 7.1). For decades, Panyu was separated from the urban centre of Guangzhou by the Ruoxi River, a major branch of the Pearl River system. The road transportation link between Panyu and Guangzhou, interrupted by the river, was connected by a ferry with limited moving facilities. Visitors to Panyu in the early 1980s frequently found cars, buses, and trucks lining up at both ends of the ferry route, with frustrated drivers and passengers complaining about the inefficient means of transportation. Although Panyu was a suburb county immediately adjacent to Guangzhou, no more than fifteen kilometres away, it normally took more than two hours to travel from the bus terminal of Guangzhou to the county capital of Panyu. In recognition of the fact that the Ruoxi River is indeed a natural obstacle hindering the free movement of goods and people between Panyu and Guangzhou, local officials decided to build a bridge to connect the two sides of the river. The project started in 1985, the year investment in the transport sector began to increase rapidly. A total of 91 million yuan, mostly mobilized through bank loans, was invested for the construction. The Ruoxi Bridge, which is 1,916 metres long with sufficient space for four lanes (15.5 metres wide), was completed three years later in August 1988. It was designed as a toll bridge so that the fee collected from vehicles could be used to pay back bank loans for its construction. The building of the Ruoxi Bridge significantly reduced the travel time between Panyu and Guangzhou from two hours to less than one hour. Such a time-space convergence has many economic and spatial consequences, discussed in the next section.

A second project of transport construction has been even more instrumental to the creation of a transactional environment for not only Panyu but also the entire Pearl River Delta region. This project is the establishment of Humen Ferry and eventually Humen Bridge in the southern end of Panyu (see Map 7.1). As discussed, Panyu is potentially a transportation hub linking much of the hinterland in the west wing of the delta region directly with the Shenzhen Special Economic Zone and the export outlet

Map 7.1

Key transport projects in Panyu, 1986-90

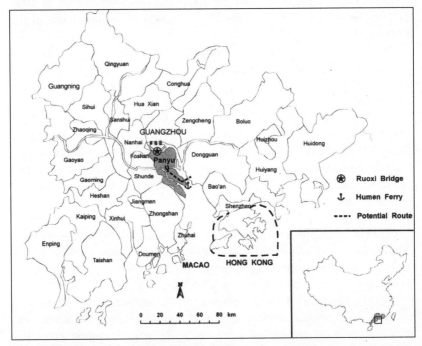

Source: Derived from field survey conducted in 1992.

of Hong Kong. This potential was recognized by local economic planners who had been exploring since the mid-1980s the possibility of bridging the two wings of the delta at Panyu. After some consultations with civil engineers at the provincial level, it was decided that a ferry be established at the southeast corner of Panyu to connect the county with Humen zhen of Dongguan (see Map 7.1). The project was funded jointly by Panyu and three Hong Kong companies, including millionaire Henry Fok, a country-man of Panyu. It started on 8 August 1989 and was completed in May 1991 (*Panyubao* 1992, June 26).

The establishment of the Humen Ferry, complemented by several dozen newly built bridges along major highways, has fundamentally improved the accessibility of Panyu for interregional transportation. Almost imme-diately after its opening, The Humen Ferry began to draw much inter-regional traffic away from Guangzhou, because the ferry provides a short cut to those who travel between the west wing of the delta and Hong Kong. The opening of the Humen Ferry has substantially reduced, by 178 kilometres, the distance between Hong Kong and such places as Jiangmen, Xinhui, and Shunde (Panyu, Urban Planning Section 1992).

While the building of the Ruoxi Bridge has enabled the people of Panyu to overcome the friction of distance between the county and the urban centre of Guangzhou, the establishment of the Humen Ferry has resulted in a spatial convergence between Hong Kong and its hinterland of the delta region.

The opening of the Humen Ferry has further stirred up a great demand for interregional transportation that is beyond the ability of the local Panyu government to satisfy. In 1992, an average of 10,000 vehicles reportedly crowded onto the Humen Ferry every day, far exceeding the ferry's daily carrying capacity of 3,000 vehicles. The demand was so high that the provincial government of Guangdong had to step in to explore the possibility of replacing the ferry with a long bridge to connect both sides of the mouth of the Pearl River. It was finally announced by the governor of Guangdong on 27 May 1992 that a bridge by the name of Humen would be built between Nansa zhen of Panyu and Humen zhen of Dongguan. A total of 1.6 billion yuan (US $334.7 million) was mobilized to fund this megaproject. The bridge is 16 kilometres long, 60 metres high, 6 lanes wide, and connected with freeways at both ends. These specifications make it the largest bridge ever built in the province, with a daily carrying capacity of up to 100,000 vehicles (*Panyubao* 1992, May 29).

The development of the transportation infrastructure has necessarily involved a substantial amount of capital investment, because many construction projects are capital intensive and their short-term economic return cannot be anticipated. Therefore, it is important to examine how such a sizeable amount of capital was mobilized and who was responsible for funding so many transport projects. Available statistical data for local public expenditure provide significant insights into how the capital of various sources contributed to recent infrastructure development.

In Chinese statistics, investment in the transport sector is usually classified under 'fixed-asset investment' (*gudin jechan touje*), because such 'fixed assets' can hardly be removed after construction. In a manner similar to production, fixed-asset investment in China normally consists of three segments, namely, public, collective, and private sectors. Although the collective and private sectors have become increasingly important in both production and investment since the reforms in 1978, the current investment structure in Panyu remains dominated by the public sector (*chuenmin shuoyiaozhi*), because almost all transport facilities or fixed assets, such as railways, highways, ports, bridges, and ferries, are constitutionally owned by the 'whole people.' Data have shown that, to all fixed-asset investment in Panyu in 1991, the public sector contributed 71.5 percent, more than the combined collective and private sectors (Guangzhou, Statistical Bureau 1992:162).

Does the dominance of the public sector in the investment of fixed assets mean that the state or central government provided most of the capital? A close examination of the sources of fixed-asset capital reveals that the state has actually contributed little to the formation of fixed-asset capital. As shown in Figure 7.2, the amount of capital made available through the state's budgetary expenditure accounted for only 0.3 percent of the total capital from the public sector. Most of the capital was mobilized through bank loans (41 percent) and county-level budgetary expenditure (27 percent). Foreign capital has also contributed a significant 16 percent of the total capital. This pattern of capital formation clearly suggests that the chief funding agent of fixed-asset investment is not the state or central government. Instead, it is the local government that plays a leading role in mobilizing a substantial amount of capital through various channels. Consequently, the construction of the transportation infrastructure in Panyu appears to owe more to local initiatives than to the state, in spite of the fact that most transportation facilities are constitutionally public or state owned.

Economic and Spatial Consequences

The motives driving Panyu's economic planners to invest heavily in the transportation infrastructure were, obviously, to spark the engine of economic growth, to encourage greater overseas investment, and to pull Panyu out of stagnant growth. To what extent have these goals of development been fulfilled? How significant has the development of the transportation

Figure 7.2

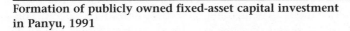

Formation of publicly owned fixed-asset capital investment in Panyu, 1991

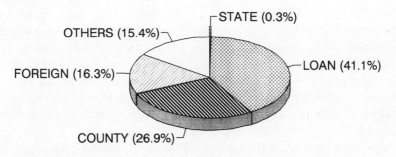

Note: The five identified categories represent the following sources of funding:
STATE: budgetary allocation from central, provincial, and municipal governments.
LOAN: loans obtained by the local governments from banks in China.
COUNTY: capital mobilized from local sources.
FOREIGN: capital obtained from foreign countries, Hong Kong, Taiwan, and Macao.
OTHERS: other funding sources.

Source: Guangzhou, Statistical Bureau 1992:199.

infrastructure been in facilitating the transformation of the spatial economy? As discussed, heavy investment in the transportation infrastructure did not occur until 1985, when capital investment in the transport sector recorded an unprecedented net increase of 37.6 million yuan in one year, and the rank of the transport sector in the local budget allocation increased from sixth to second position (Guangzhou, Statistical Bureau 1989:395). Most of the investment was directed toward the construction of bridges and highways. It is thus not surprising to find that the mileage of highways was substantially extended by 199 kilometres in 1985 alone, more than double the total length of all highways constructed in Panyu's history (see Figure 7.3). In the case of bridge construction, the dramatic increase in the number of bridges occurred in 1986, but the construction of new bridges actually took place one year earlier in 1985. Thus, 1985 was a turning point for both investment in the transport sector and construction of transport facilities. After 1985, transport development began to emerge as a remarkable force shaping Panyu's spatial economy (see Figure 7.3).

With the time frame for transport development clarified, we can now assess how and to what extent such development has affected the process of spatial transformation. As 1985 marked the beginning of heavy investment in the transportation infrastructure, one possible strategy is to compare economic growth before and after 1985, determining if transport development has genuinely facilitated the growth of the local economy. The most common index for economic growth in Chinese statistics is the Gross Value of Industrial and Agricultural Output (GVIAO). Data have shown that industrial and agricultural output value grew at an annual rate of 15.94 percent from 1980 to 1985, before heavy investment in the transport sector was initiated (Guangdong, Statistical Bureau 1992b:99). This growth rate was significantly lower than those of Shunde, Nanhai, Zhongshan, and Dongguan, which were traditionally counterparts of Panyu (see Figure 7.4). It was even lower than the regional average of the Pearl River Delta, which recorded an annual 16.75 percent increase in GVIAO for the same period (Guangdong, Statistical Bureau 1992b:65). Obviously, the economic performance of Panyu was less than ideal before 1985. It should be stressed that this pattern of slow economic growth existed in spite of the fact that economic reforms and the open door policy had been implemented in 1979.

When Panyu entered the second half of the 1980s, its production of industrial and agricultural output increased at an unprecedented annual rate of 32.38 percent, which was more than double its growth rate in the previous period (Guangdong, Statistical Bureau 1992b:99). Whereas Panyu had lagged behind many of its rivals in economic growth during the previous period, it now jumped to the forerunner's position. Panyu led not only the delta region as a whole, which was growing at an aver-

Figure 7.3

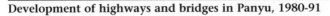

Development of highways and bridges in Panyu, 1980-91

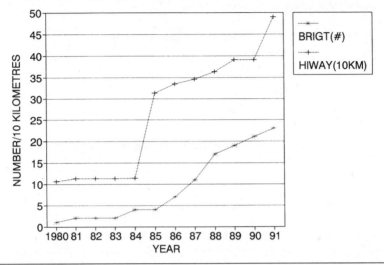

Note: BRIGT (#) stands for the number of bridges and HIWAY (10KM) for the total length of highways in ten kilometres.

Source: Panyu, Statistical Bureau 1989:36; 1992:71.

age annual rate of 29 percent, but also some of the economically advanced counties such as Shunde, Nanhai, and Zhongshan (see Figure 7.5). While the very rapid growth of industrial and agricultural production might be attributed to many social, economic, and geographical factors, the fact remains that the fundamental improvement in the transportation infrastructure, which featured the development strategy adopted by local economic planners since 1985, significantly contributed to the transformation of Panyu from 'a land of standstill' to a place of accelerated development. The positive relationship between transport development and economic growth was also supported by statistical analysis using Pearson correlation coefficients that shows significant relationships between the growth of GVIAO and the increase of transport investment ($r = 0.94$), and the number of bridges built ($r = 0.89$), as well as the mileage of highways extended ($r = 0.82$).

It appears that transport development was a crucial driving force, albeit not the only one, that facilitated the transformation of the local economy. But in what manner has transport development interacted with economic growth? Does it occur before, after, or simply concurrently with the expansion of the local economy? To better understand the process of development in Panyu, statistical data for the gross value of industrial and agricultural output (GVIAO), the number of bridges constructed (BRIGT), and the

Figure 7.4

Annual growth of GVIAO for Panyu in comparison with other economically advanced places, 1980-5

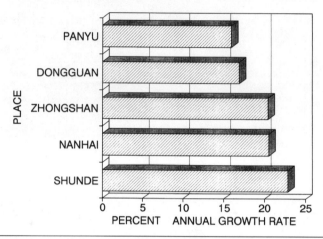

Note: GVIAO stands for Gross Value of Industrial and Agricultural Output. Raw data are at 1980 constant prices.

Source: Guangdong, Statistical Bureau 1992b:99, 135, 139, 171, 175.

Figure 7.5

Annual growth of GVIAO for Panyu in comparison with other economically advanced places, 1985-91

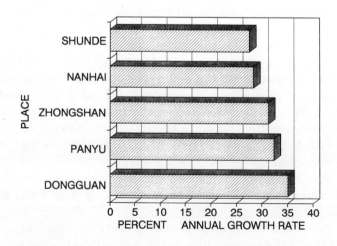

Note: GVIAO stands for Gross Value of Industrial and Agricultural Output. Raw data are at 1980 constant prices.

Source: Guangdong, Statistical Bureau 1992b:99, 135, 139, 171, 175.

mileage of highway extended (HIWAY) are plotted in Figure 7.6 for analysis. The variable of GVIAO was selected because it is the most common index for economic growth in China for which historically comparable data are available. The other two selected variables represent the construction of bridges and highways, the focus of transport development since the mid-1980s. The resulting picture of growth suggests that the expansion of the local economy in terms of industrial and agricultural production has been generally consistent with the improvement in transportation facilities, particularly highways and bridges. However, the pace of change in production has not fit neatly with the increase in the number of bridges or the extension of highways. A closer examination of the process of growth reveals that a lag relationship exists between the construction of transport facilities and the expansion of industrial and agricultural production. As shown in Figure 7.6, a turning point for highway extension was clearly marked in 1985. A significant increase in the number of bridges occurred in 1986. It should be noted, however, that construction for most of the new bridges took place at least one year earlier. This pattern of growth is consistent with that of the increase of investment in the transport sector, which started in 1985 (see Figure 7.1). While heavy investment in the transport sector and large-scale construction of highways and bridges took

Figure 7.6

Growth of GVIAO and transport development in Panyu, 1980-91

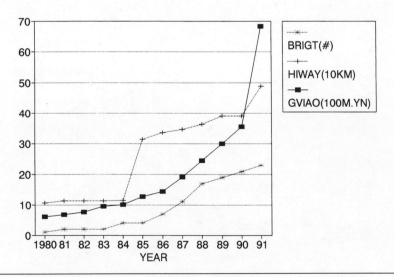

Note: BRIGT (#) stands for the number of bridges, HIWAY (10KM) for the total length of highways in ten kilometres, and GVIAO (100M.YN) for Gross Value of Industrial and Agricultural Output in 100 million yuan at 1980 constant prices

Sources: Panyu, Statistical Bureau 1989:36; 1992:71; Guangdong, Statistical Bureau 1991b:30,32.

place in 1985, industrial and agricultural production did not make a remarkable increase until 1987. In other words, the turning point for economic growth lagged by two years.

A similar lag relationship exists between transport development and the growth of foreign investment as well as export production. As shown in Figure 7.7, foreign investment and export production were stagnant during the first half of the 1980s, in spite of the fact that the open door policy was implemented in 1979. It was not until 1986 that both foreign investment and export production started to move upward. The first turning point for the growth of both foreign investment and export production was clearly marked in 1986, a year after massive transport construction commenced.

The lag relationship between transport development and economic growth has significant theoretical implications. Since the 1970s, assessment of the economic impact of transport development has emphasized its facilitative role, in which the expansion of transportation facilities is seen as a subsequent response to economic needs rather than as an independent force that will induce new economic activities (Gauthier 1970:613; Taaffe and Gauthier 1973:200; Hoyle and Hilling 1984:4;

Figure 7.7

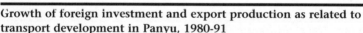

Growth of foreign investment and export production as related to transport development in Panyu, 1980-91

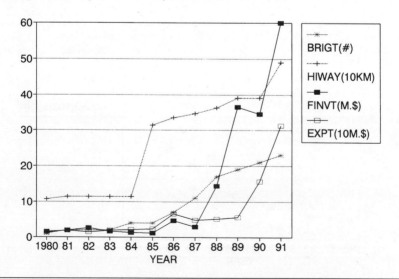

Note: BRIGT (#) stands for the number of bridges, HIWAY (10KM) for the total length of highways in ten kilometres, FINVT (M.$) for realized foreign investment in US $million, and EXPT (10M.$) for export output value in US $10 million.

Source: Panyu, Statistical Bureau 1989:36; 1992:71; Guangdong, Statistical Bureau 1991b:32.

Leinbach and Chia 1989:3). This perspective on the role of transport development may be valid for many economically advanced countries, but it may not necessarily hold true for Panyu, where economic growth is still at an early stage. The experience of Panyu, where economic development has long been impeded by the lack of an efficient road transportation network, tends to suggest that the expansion of transport facilities occurred before rather than after the dramatic growth of the local economy. It is the improvement in the transportation infrastructure that created a transactional environment conducive to foreign investment, increased exports, and greater manufacturing and agricultural output.

The dramatic growth of the local economy since the mid-1980s as a result of improvement in transport facilities has also provided an explanation for the distinctive feature of economic growth of the Pearl River Delta, where the increase in economic production of the suburban counties has outstripped that of the central cities described in Chapter 5. With its suburban location adjacent to the central city of Guangzhou, Panyu is an ideal place not only for industrial distribution from Guangzhou but also for manufacturing branches subordinate to the major plants in the central city. This potential geographical advantage was not, however, fully exploited before 1985, because transportation linkages between Panyu and Guangzhou were inefficient. With the improvement in the road transportation infrastructure, particularly the construction of Ruoxi Bridge, which substantially reduced the travelling time between Panyu and Guangzhou, it became feasible for Panyu to supply industrial materials needed in the central city. Such a development was reinforced by the fact that manufacturing expansion in Guangzhou had been constrained by urban congestion problems as well as environmental pollution regulations (Lin 1986).

A great number of factories have been set up in Panyu to produce manufactured parts for assembly in Guangzhou. This development is especially evident in Dashi zhen, located right on the border between Guangzhou and Panyu. Dashi formerly was separated from Guangzhou by the Ruoxi River. After the Ruoxi Bridge was built in 1988, Dashi suddenly emerged as a satellite town for Guangzhou. Many factories were established as subbranches of modern industries in Guangzhou to produce a variety of manufactured materials, including parts for refrigerators, computers, cameras, electric fans, and washing machines. As a result, the output value of manufacturing shot up dramatically from 70.67 million yuan in 1988 to 1.18 billion yuan in 1990, a net increase of 47.09 million yuan or 67 percent in two years (Panyu 1992, interview). This pattern of growth suggests that the improved efficiency of the transportation network of a suburban county such as Panyu is one crucial factor that has contributed to growth in the Pearl River Delta, primarily characterized by an accelerated expansion of economic production in the suburban areas.

Improvement in the road transportation infrastructure has also signifi-cantly increased the accessibility of the county and induced a growing volume of in-migration. With its proximity to and improved transport connections with Guangzhou, Panyu since the mid-1980s has become a favoured destination for migrants from the large city of Guangzhou, where new immigrants could hardly stay because of urban congestion and the official restriction on migration to the cities. Many new immigrants were from the less-developed interior provinces such as Sichuan, Hunan, and Anhui. Most of them wanted to become employed in Guangzhou. After they arrived in the city, however, they realized that there was no room for them to stay, given the constant threats of deportation from the city authorities and the difficulty in finding accommodation in the large city. Consequently, many of these immigrants were brought by their coun-tryfolk through an incredibly well-organized underground network extending from the railway station of Guangzhou to nearby suburban counties such as Panyu. A majority of these immigrants were young women between the ages of eighteen and twenty-five. Through the underground networks operated by their countryfolk, newcomers were introduced to pos-sible employers and began to work as factory workers, babysitters, wait-resses, secretaries, or even prostitutes. Statistically, these new immigrants are called 'temporary population,' lacking official permanent resident status but residing in Panyu for at least ten months.

Data obtained from the Public Security Bureau of Panyu show that the temporary population never exceeded 5,000 before 1986. Since then, it has dramatically grown, first to 15,741 in 1986, and then 89,167 in 1991. While the large increase in migration could be attributed to a variety of factors, including the relaxation of government control on the rural-urban migration effective in 1984, the improvement of road transport facilities is unquestionably a critical force that has significantly facilitated migration to and within Panyu. The advance in transportation facilities has thus become another important factor that explains why the dramatic growth of population, particularly migration, has primarily occurred in suburban counties such as Panyu rather than the central city of Guangzhou, a dis-tinctive demographic feature of metropolitan development in the Pearl River Delta identified in Chapter 5.

The growth of migration to and within Panyu demonstrates a lag rela-tionship with transport development. As shown in Figure 7.8, the number of temporary population had not experienced any significant growth from 1980 to 1985. Its dramatic increase occurred first in 1986, one year after massive transport development and construction were initiated. This pat-tern of growth is consistent with those of foreign investment, export, and economic production outlined in the previous section. The creation of a more accessible and efficient road transportation network has thus not only quickened the pace of economic growth but also fostered the mobility of population in the area.

In addition to economic growth and migration, transport development has also facilitated the land-use transformation that has been taking place mainly along major transportation arteries. Although detailed data on land-use change were not available to allow a systematic assessment, my field survey of recent development in Panyu has nevertheless uncovered some significant patterns of land-use change that are associated directly or indirectly with the improvement of the local transportation network.

In a manner similar to the growth of industries relocated from Guangzhou, the improved transportation linkages between Panyu and Guangzhou have encouraged many suburban activities to take place in Panyu. Since the 1988 completion of Ruoxi Bridge, many villa-style apartment buildings, shopping malls, restaurants, and recreation centres have emerged in Panyu, primarily to meet the needs of the people of Guangzhou who have suffered from urban congestion in the central city, and who have been looking for a more relaxed outlet for accommoda-

Figure 7.8

Growth of temporary population and transport development in Panyu, 1980-91

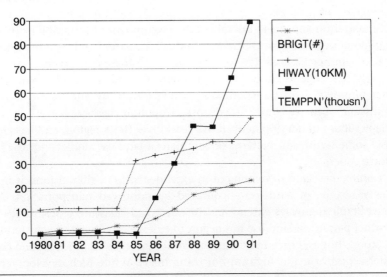

Note: Temporary population refers to those migrants who have registered with the local authority and resided in Panyu for ten months or longer. BRIGT (#) stands for the number of bridges, HIWAY (10KM) for the total length of highways in ten kilometres, and TEMPPN'(thousn') for the temporary population in thousands.

Sources: Panyu, Statistical Bureau 1989:36; 1992:71; Panyu, Public Security Bureau, Archive Office, 1992, unpublished annual reports for internal use.

tion, leisure, and recreation. Such suburban developments have mixed locations, but most of them are concentrated in the corridor between Guangzhou and Shiqiao zhen, which is the county capital of Panyu.

One distinctive type of land use that recently emerged as a result of transport development is an extended strip or ribbon of specialized stores selling the same type of goods. On the road from Guangzhou to the county capital of Panyu, for example, are hundreds of furniture stores, a ribbon starting at the southern end of the Ruoxi Bridge and extending for many miles. This area used to be primarily rice fields. It was not until 1988, when road transportation between Panyu and Guangzhou was made possible by the Ruoxi Bridge, that commercial land use began to take over. Most of the furniture stores were developed to attract consumers from Guangzhou, located just on the other side of the Ruoxi Bridge. The prices for furniture are adjusted according to the working schedules of the people of Guangzhou. During weekdays, when demand for furniture is low because most people in Guangzhou are at work, furniture is sold at a 20 percent discount. On holidays and weekends, when the number of shoppers from Guangzhou is greater, the prices for furniture are raised to their usual high level.

The idea of grouping furniture stores along major traffic roads is clearly to take full advantage of convenient transportation and to attract consumers by offering a maximum number of choices for shopping. The emergence of a ribbon of specialized stores as a result of improved road transportation appears to resemble the development of shopping malls in the suburbs of many North American cities. The extent and pattern of suburban commercial land development in the Chinese context may not be exactly the same as those in North America. However, the fact remains that improved road transportation, which effectively altered the urban landscape of many North American metropolitan regions, has started to shape suburban development in the Pearl River Delta region, and has created some significant patterns of commercial land use similar to those in North America.

Commercial land development in extended strip or ribbon form is not the only type of land use encouraged by improved transportation. As newly built highways extend and intersect with existing roads, another distinct pattern of land use has begun to emerge, involving the clustering of stores, hotels, office buildings, and other business facilities around the intersection of major highways or trunk roads. While such development may be beneficial to doing business, it has mixed the interregional movement of through-traffic and the movement of pedestrians, causing much traffic congestion. The local solution to this problem has been to construct a flyover or 'pedestrian crossing bridge' (*renxin tienqiao*) at an intersection so that local pedestrians can walk around and do their shopping safely while interregional traffic passes by. Such a nodal form of land-use development, primarily driven by the extension and intersection of highways, has become a distinctive feature of landscape not only in Panyu but also in other cities and towns of the Pearl River Delta region.

The impact of transport development on land-use change has also been felt in the countryside and even in some remote areas. On the southern frontier of Panyu, where road transportation is only a recent phenomenon, much farmland has been encroached upon by highway construction and the subsequent industrial and commercial developments taking place on both sides of newly built highways. The spatial outcome of this land-use transformation has been the emergence of a number of discrete corridors, where restaurants, hotels, gas stations, garages, and drug stores are built along extended highways and near the entrances of villages or towns. These discrete corridors, which are shaped by highway frontier extension, may later evolve into a large-scale land development in ribbon or nodal forms as described earlier.

Other land-use developments directly or indirectly related to the improved transport network have been the establishment of Nansa zhen near the Humen Ferry as an Economic and Technological Development Zone, the 1992 expansion of the urban area of the county capital from a mere 10 square kilometres to 41 square kilometres (*Panyubao* 1992, January 3), and the construction of numerous villa-style apartment buildings and luxurious hotels to attract wealthy people from Guangzhou and Hong Kong. These land developments, encouraged directly or indirectly by the improvement of the transport network, have contributed to the dramatic reduction of farmland that has been significantly accelerated since the mid-1980s. In 1989, for instance, a total of 14,303 mu or 2,355 acres of cultivated land was lost to industrial, transport, commercial, and residential uses (Panyu, Land Development Section 1992). While transport development may not be the only force responsible for all the changes of land use that have occurred, it is nonetheless a critical factor in facilitating the process of land-use transformation.

Ironically, the construction of bridges and highways has not improved traffic mobility. Instead, it has exacerbated the problems of traffic congestion and traffic accidents as more interregional traffic has been channelled into Panyu. It has been reported by the Public Security Bureau of Panyu that the number of traffic accidents has increased rapidly in recent years. Between 1 January and 31 August 1992, for instance, a total of 106 casualties caused by traffic accidents was reported. In August alone, 52 reported traffic accidents claimed 22 lives. As the road transportation network has improved, the number of vehicles, especially motorcycles, has skyrocketed. Data obtained from the local Public Security Bureau indicate that the number of registered motorcycles owned by the people of Panyu had drastically increased from a mere 85 in 1978 to 25,056 in 1991. Consequently, visitors to Panyu have frequently found themselves overwhelmed by numerous motorcycles sharing a lane with other motor vehicles. Development of the transportation infrastructure of Panyu has, therefore, not only facilitated economic growth, migration, and land-use changes, but also created a greater demand for further expansion of the existing road transport network.

Summary

In his seminal work on the emergence of the megalopolis in the United States, Jean Gottmann highlighted modern transportation and telecommunication as an essential condition for the mass movement of population and commodities in the urbanized region (Gottmann 1961:632). Thirty years later, similar efforts were made by McGee and Ginsburg, who drew our attention to the process of settlement transition in Asia, where a low-tech 'transportation revolution' had begun to shape extended metropolitan regions (McGee 1991b; Ginsburg 1990). Although the spatial effect of transport development has long been a subject for scholarly contemplation, little empirical work has been done to show how transportation has in fact interacted with the transformation of a regional economy.

This chapter assesses the role played by transport development in the process of spatial transformation, using Panyu as a case study. The results of data analysis and field survey have suggested that heavy investment in transport infrastructure has created a transactional environment conducive to economic growth, migration, and land-use transformation. The creation of a more efficient road transportation network integrated by bridges, highways, and ferries has significantly improved the accessibility of Panyu, strengthened its economic linkages with the central city of Guangzhou, and encouraged suburban industrial and commercial activities in this previously forgotten 'land of standstill.' Overcoming the 'friction of distance' existing between Panyu and Guangzhou has been a major focus of the development strategy adopted by local economic planners. It has also become a factor of vital importance in understanding why dramatic economic and spatial changes could have occurred in this suburban county of the Guangzhou metropolitan region.

For a place like Panyu, whose economic growth was severely blocked because of its inefficient transport network, the provision of a well-established road infrastructure as a necessary, albeit not sufficient, condition has proved to have the effect of leading to rather than following the transformation of the local economy. Economic growth in Panyu in terms of production, export, and foreign investment did not accelerate until after the mid-1980s, when massive construction of bridges and highways took place. The extension of a road network has effectively altered the suburban landscape and created distinct patterns of land use such as discrete corridors and commercial development in ribbon and nodal forms. The experience of Panyu, where economic growth is still in an early stage, has appeared to differ significantly from that in countries where development has already reached a mature stage, and for which transport expansion is normally perceived as a passive response to increased economic demand.

The development of the transportation infrastructure has been funded primarily by the local government. The central state has not played any active role in planning or funding the improvement of local transport facilities. With the demise of a centrally planned economy, a new development mechanism seems to have taken shape in which initiatives are made primarily by local governments responsible for mobilizing necessary resources, setting up modern transport infrastructure, and leading the local community toward economic prosperity.

8
Influence of Hong Kong

In the current age of the internationalization of investment and production, the transformation of a spatial economy in China has increasingly found itself being shaped by global forces emanating from the restructuring capitalist world. The structural and spatial effects of global market forces are probably more noticeable in the Pearl River Delta region than anywhere else in the nation. As a Chinese frontier with ready access to Hong Kong and overseas, the Pearl River Delta has since 1979 been allowed to move ahead of the nation to attract foreign capital, acquire Western advanced technology, and develop a market-oriented economy. Its great openness to the outside world and intensified social and economic linkages with Hong Kong have given the Pearl River Delta a special regional identity. To understand the dynamics of growth and spatial changes in the delta, it is necessary to go beyond the previous analyses of local initiatives to examine how the intrusion of global market forces through Hong Kong has helped shape the new spatial economy.

To single out the influence of Hong Kong as an important factor for analysis is also based on the consideration that this issue has been a growing concern among scholars and that many associated structural and spatial questions remain unresolved. The economic integration of Hong Kong and Guangdong Province, as a result of implementing the open door policy, has been addressed by a number of scholars. Thus far, the bulk of the research has tended to emphasize the increasing interaction between Hong Kong and Guangdong (Vogel 1989; Sung 1991; Kwok and So 1995), including the incentives for relocating manufacturing facilities from Hong Kong to the delta (Leung 1993; Thoburn, Leung, Chau, and Tang 1990), and the future prospects for development in this region after China reclaims sovereignty over Hong Kong in 1997. Little is known, however, about the impacts of Hong Kong initiatives on the transformation of the spatial economy of the Pearl River Delta.

How and to what extent has the relocation of manufacturing from Hong Kong to the delta region contributed to the growth and restructuring of the delta's economy in terms of production and employment? What impact has Hong Kong investment had on the growth of immigration and the spatial redistribution of population? How has the transformation of land use been facilitated by the establishment of Hong Kong manufacturing firms in the region? These questions are essential to understanding the dramatic changes in the delta region.

This chapter assesses the economic and spatial consequences of the interaction between Hong Kong and the Pearl River Delta region since China's open door policy was initiated in 1979. As existing data are not detailed and systematic enough to allow for an assessment of the development of the entire region, this chapter will focus on a case study of Dongguan, a county-level municipality considered by many Chinese scholars to be a classic example of structural and spatial change resulting from the establishment of subcontracting firms in the delta region. Because the influence of Hong Kong on the delta is not limited to the economic sphere of investment and manufacturing, the social and cultural aspects of Hong Kong's influence are included in this assessment.

Geographical and Historical Context

Among the counties and cities in the Pearl River Delta, Dongguan is probably the best case to show how a regional economy can be significantly transformed by forces primarily from Hong Kong. Located in the eastern wing of the delta in proximity to Hong Kong, Dongguan is typical of the situation in the delta region and, to some extent, representative of Guangdong Province in the national context. Although Dongguan is second to Baoan and Shenzhen in terms of distance to Hong Kong, it has fared better in developing kinship ties with Hong Kong, primarily due to its higher population density and its well-established historical connections. In 1986, Dongguan residents had at least 650,000 relatives in Hong Kong and Macao, a number significantly higher than those of Baoan and Shenzhen (Guangdong, Land Development Bureau 1986:369-70). The central position of Dongguan in the Guangzhou-Hong Kong corridor means that it can easily access both the export outlet of Hong Kong and the traditional urban centre of Guangzhou. Such a geographical location has also enabled Dongguan to develop an export-processing industry by merging capital, technology, and industrial parts from Hong Kong in the south with interior cheap labour transferred mainly from Guangzhou in the northwest.

The favourable geographical features of Dongguan, however, brought no benefit to the local economy in the Maoist era when traditional connections between Hong Kong and the mainland were arbitrarily cut off. Under the then-prevailing radical ideology of anti-capitalism, the frontier position of Dongguan in relation to the capitalist territory of Hong Kong was considered vulnerable not only to the 'contamination' of capitalism but also to possible naval attacks from counter-revolutionary enemies overseas, including those in Taiwan and the United States who had constantly threatened to roll back the Chinese revolution. The perceived vulnerability of Dongguan's location to capitalist attack explains the fact that for decades Dongguan had never become the focus of infrastructure development funded by the central or the provincial government.

Local people had few alternatives to make a living, except by working in the rice fields to 'learn from Dazhai' in building a self-reliant agrarian economy. On the eve of economic reforms, a total of 390,000 persons or 72 percent of the total labour force was engaged in agricultural production (CCP Team 1989:27). Annual per capita income was a mere 193 yuan (CCP Team 1989:3). In some places, such as the southern border township of Chang'an, annual per capita income was recorded at the unbelievably low level of 83 yuan (CCP Team 1989:5). Tens of thousands of young people had to run the risk of the death penalty to escape to Hong Kong. About 20 percent of the young people in Dongguan reportedly managed to get across the border into Hong Kong in the pre-reform years, primarily because they had no hope of a reasonable future in their hometown (Vogel 1989:176).

The implementation of the open door policy led Dongguan to enter a new era of development. The opening up of Guangdong and Fujian provinces, the establishment of Special Economic Zones, and the designation of the Pearl River Delta as an Open Economic Region have renewed and strengthened the economic ties between the delta and Hong Kong. Under this new circumstance, geographical proximity to and wide-ranging personal connections with Hong Kong, which used to be detrimental to Dongguan, have now become valuable assets that can be used to attract overseas investment and develop an export-oriented economy.

Development of an Export-Processing Industry

After local people realized that their advantageous connections with Hong Kong could be used to create jobs and raise income, they began to seek every possible opportunity to promote investment from Hong Kong and overseas. Special policies, including taxation concessions and preferential treatment for the import of necessary equipment and handling of foreign currency, were announced to attract foreign investment. A special office was set up to serve Hong Kong investors with efficient personnel and simplified bureaucratic procedures. Economic cooperation between Dongguan and Hong Kong was arranged creatively and flexibly in a variety of forms including export processing (*lailiao jiagong*), compensation trade (*buchang maoyi*), joint ventures (*hezi jingying*), and cooperative ventures (*hezuo jingying*). It was reported that by the end of 1991, a total of 10,586 contracts had been signed between Dongguan and manufacturers from Hong Kong and overseas, of which 5,700 were already in operation (Guangdong, Statistical Bureau 1992b:20).

Economic cooperative ventures developed since 1979 between Dongguan and Hong Kong have varied in form and in size, but the most popular venture has been the processing of imported materials (*lailiao jiagong*) or the assembling of parts provided by Hong Kong manufacturers (*laijien chuanpei*). Known as 'three supplies, one compensation' (*sanlai yibu*), the arrangement requires Hong Kong to supply raw materials, components or parts, and models for what is to be processed, while the Chinese provide labour, land, buildings, electricity, and other local utilities necessary for production. The Hong Kong participant in the contract does not hire or pay workers directly. Instead, a lump-sum payment is usually made available to the Dongguan participant for the contracted goods. With payment from Hong Kong, usually in American or Hong Kong dollars, and paid in instalments until the products are completed, the Dongguan participant in the contract hires workers and pays them in Chinese dollars on a piecework basis. Needless to say, local governments and cadres of Dongguan, who serve as intermediaries in this process, are able to make sizeable profits by paying low salaries or by exchanging American and Hong Kong dollars into Chinese yuan at a high rate on the black market. Arrangements are also made on a compensational basis, where the Dongguan partner processes or assembles jobs for a set time, such as five years, and then assumes ownership of the machinery or equipment provided by the Hong Kong firm as compensation.

Cooperation in the form of 'three supplies, one compensation' has become popular not only in Dongguan but also in other parts of the delta region, because it has substantially benefited both Hong Kong and Chinese participants in the contract. With designing and marketing handled in Hong Kong and labour-intensive work done cheaply in Dongguan, small Hong Kong manufacturers are able to compete well in the international market. As for the Chinese side, export processing has created jobs and income for local cadres and the general population. By the end of 1987, some 2,500 processing firms on the basis of 'three supplies, one compensation' had been set up in Dongguan, creating up to 171,000 employment opportunities, and receiving US$107 million, mostly from Hong Kong, which accounted for about 40 percent of the total received by Guangdong Province (CCP Team 1989:6). By the end of 1990, the number of export-processing firms established in Dongguan had reached 4,680. These firms produced a total export output of US$150.06 million, more than all other cities and counties in the province except the Shenzhen Special Economic Zone (Guangdong, Statistical Bureau 1991c:357). A survey conducted by the Federation of Hong Kong Manufacturers in July 1991 identified Dongguan as the second most favoured location, next only to Shenzhen, for Hong Kong investment (Hong Kong, Federation of Hong Kong Industries 1992:13). The considerable success of Dongguan in attracting foreign investment and developing export manufacturing resulted in its promotion from a county to an officially designated municipality at the county level in 1985, and to a higher-level municipality directly subordinate to the provincial government in 1988.

Reasons for Developing an Export-Processing Industry

Why has Dongguan, formerly a frontier agrarian county, become favoured by Hong Kong manufacturers? What are the forces that have helped Dongguan to attract investment and processing activities from Hong Kong and overseas? In answering these questions, local cadres have frequently quoted the words of well-known ancient Chinese scholar and strategist Zhuge Liang that 'timing, location, and public relations' (*tianshi, dili, renhe*) are the three essential factors in seeking any success. Implied in this explanation is the implementation of the open door policy (timing), the geographical proximity of Dongguan to Hong Kong (location), and the creation of good personal relations with Hong Kong investors (public relations). While the case of Dongguan appears to fit fairly well into the general model of success provided by the ancient Chinese strategist, three specific factors should be highlighted to understand the rapid expansion of export-processing activities.

First, the good personal connections between Dongguan and Hong Kong have provided easy and reliable links between investors and their manufacturing partners. With over 650,000 relatives (*gang'ao tongbao*) in Hong Kong and another 180,000 (*huaqiao*) in other foreign countries, mostly in North America, people here have less difficulty than those in other parts of China in seeking investors or partners from Hong Kong and overseas. It was estimated by local cadres that about half of the contracts they had signed were with former Dongguan residents now living in Hong Kong. Interestingly, many personal contacts are with those who escaped from Dongguan to Hong Kong during the pre-reform period. Ironically, local cadres who used to be responsible for preventing escapes and apprehending those who dared to try are now in charge of contacting and persuading escapees in Hong Kong to invest in their native county (Vogel 1989:176).

A second critical contributing factor, often overlooked in the assessment of the growth of export processing, is the creation of a transportation infrastructure as a necessary means to attract foreign investment. In this regard, the local government has played a leading role in the development process. It was reported that from 1980 to 1987 a total of 1.034 billion yuan (US$216 million) was raised by the local government through various channels for infrastructure development (CCP Team 1989:39). Such a huge amount of capital was obtained primarily from local resources such as bank loans (33 percent), collective enterprises (31 percent), stocks and bonds (14 percent), and foreign capital (11 percent). Budgetary allocations from the central and provincial governments accounted for only 11 percent of all construction expense (CCP Team 1989:35).

Heavy investment in the infrastructure has resulted in significant improvements. The existing road system has been substantially extended with the mileage of paved roads increasing from a single kilometre in 1978 to 860 kilometres in 1987. By the end of the 1980s, Dongguan had more miles of paved roads per square kilometre than any other county in the nation (CCP Team 1989:7). Dongguan was also one of the first Chinese counties to establish a computerized telephone system, connecting it directly with seventeen countries and regions in the world. A total of 13,231 telephones were installed in the townships and villages in 1987, of which 8,756 phones, or 20 percent of all installations in China, could dial direct to other countries (CCP Team 1989:7, 34, 37). The transport capacity of ports and harbours and the generation of electric power also increased substantially during the 1980s. The creation of such a good infrastructure has significantly reduced transactional costs for investors and, therefore, underpinned the rapid inflow of overseas investment.

Finally, the availability of cheap labour and land is another important factor that has helped attract Hong Kong manufacturing to move into Dongguan. In the early 1980s, Dongguan was a county where labour and land could be obtained easily and cheaply. A worker employed by an export-processing firm was usually paid a monthly wage of 150 yuan to 200 yuan, about one-fifth or even one-sixth of what a Hong Kong worker could make (CCP Team 1989:194). Although Chinese workers may not be as skilful as their Hong Kong counterparts in certain industrial productions, the low wages remain attractive to Hong Kong manufacturers, especially to those who are engaged in highly labour-intensive industries such as toys and electronics. Since 1979, the increase in employment opportunities has resulted in a tendency toward rising labour costs. This tendency, however, has been balanced by an influx of new workers from less-developed interior provinces who will accept low wages. Consequently, low labour costs in Dongguan remain a significant factor that continues to attract manufacturing from Hong Kong and overseas (CCP Team 1989:194).

Characteristics of the Export-Processing Industry
As the export-processing industry continues to grow, it is becoming one of the most dynamic economic sectors in transforming the regional economy

of Dongguan. By the end of 1990, more than 70 percent of Dongguan's industrial labour force was engaged in export processing (*Yatai Jingji Shibao* 1992, August 2). The continuing expansion of the industry has led the local economy to enter a new stage of accelerated growth and restructuring. It has also effectively altered the economic landscape of Dongguan and created some distinct spatial patterns associated with industrial production, migration, and land use. As the transformation of the local spatial economy was to a great extent fuelled by the relocation of manufacturing from Hong Kong, to understand the dynamics of these structural and spatial changes requires an analysis of the nature and spatial characteristics of the flourishing export-processing activities.

Hong Kong manufacturers relocated their workshops from Hong Kong to Dongguan primarily to tap the existing pool of low-priced inexperienced labour. It is, therefore, not surprising to find that resulting industries are simple, unsophisticated, small scale, and labour intensive. In the main, export processing in Dongguan has centred on four sectors: textile, apparel, toys, and electronics. The type of production varies considerably, from processing toys, assembling simple radios, and sewing shirts or blouses, to making plastic bags, incense, firecrackers, candles, candy, chocolate, and other food products. The procedures of production are invariably simple and repetitive, requiring a considerable amount of time and labour but little skill. The growth of these simple labour-intensive industries has significant implications for changes in the employment structure and migration as it opens up opportunities for those surplus rural labourers who are eager to enter factories but have little experience or skill in manufacturing.

As processing activities are technologically unsophisticated, many factories that have been set up are relatively small in size. Most of them do not require heavy machinery. Some were converted from the dining halls of former communes or brigades. As production expands, buildings of two or three stories are constructed containing several large rooms which accommodate 50 to 100 desks, one for each worker. Thus, a typical factory may employ several dozen to 100 workers, which is small by Chinese standards. A 1991 survey sampling 2,931 joint ventures and compensational trade enterprises in Dongguan revealed that the average number of workers in each factory was 147 for joint ventures and 105 for compensation trade factories (Lu 1992:146). Some workshops in the countryside were so small that they had only a dozen workers on their payrolls. The fact that the export-processing industry consists of numerous small workshops without a single major plant has been vividly described by the local people as 'a spread of numerous stars in the sky without a large shining moon in the centre' (*mantian xingdou queshao yilun mingyue*).

Another closely related feature of Dongguan's export-processing industry is that the factories are not concentrated in a few large urban centres but are widely scattered throughout the countryside. Since the scale of production is small and the processing procedure is simple, factories in Dongguan do not necessarily have to be located in the large urban centres where technical experts or high-ranking social services are easily accessible. Rather, existing personal kinship ties, the improved transport and electric power infrastructure, an abundant supply of cheap surplus rural labour and land space, and a less-regulated environment have all combined to attract investment and manufacturing activities from Hong Kong to the rural townships and villages. This distinctive feature is evident from an official survey conducted at the end of 1987, which revealed that Dongguan's export-processing factories were predominantly located in the townships and villages of the countryside. Of the 2,500 factories established for the processing of imported materials or compensation trade, 1,591 were in rural townships and villages. They accounted for 63.64 percent of the total number of export-processing firms, 72.52 percent of all processing fees received from Hong Kong and overseas, and 87.91 percent of the total construction area of all factories set up for the processing industry (CCP Team 1989:32). This spatial pattern in which export-processing activities are widely distributed all over the countryside has significant implications for the process of land-use transformation, which will be discussed next.

Consequences of Export Industrial Development
Having identified the distinctive features of the nature, size, and spatial distribution of the export-processing industry, it is now possible to analyze the structural and spatial consequences of this externally driven industrial development. The most significant outcome of the flourishing of labour-intensive export-processing activities has been a disproportionate increase in employment and production in the manufacturing sector and the subsequent restructuring of the local economy. When Dongguan first opened up to foreign investment in the late 1970s, its economy was predominantly agricultural, with two-thirds of its population working in the fields at a subsistence level. The rapid expansion of the export-processing industry since 1978 has greatly increased the pace of manufacturing development. Between 1978 and 1991, an estimated 380,000 jobs were created by the export-processing industry and absorbed both local rural labourers, who were released from agricultural production, and immigrants, who moved in from less-developed areas (*Yatai Jingji Shibao* 1992, August 2). Consequently, the labour force in the secondary sector, primarily manufacturing in nature, has expanded at 10.45 percent per annum since 1978, with its share of the total labour force increasing from 16.85 percent in 1978 to 40.64 percent in 1990 (Dongguan, Statistical Bureau 1991:6). At

the same time, those engaged in agricultural production and other primary activities have become fewer in number, and their share of the total labour force has dropped substantially from 71.57 percent in 1978 to 36.15 percent in 1990 (Dongguan, Statistical Bureau 1991:6). The growth of output value exhibited a pattern similar to that of the labour force. The contribution of manufacturing to the total output value rose from 42.06 percent in 1978 to 66.20 percent in 1990, while the share of the agricultural sector declined from 39.40 percent to only 19 percent in the same period (Dongguan, Statistical Bureau 1991:20).

In addition to economic restructuring, the export-processing industry has contributed to the accelerated growth of the local economy and helped to raise personal income for the general population. From 1980 to 1990, the growth of industrial and agricultural output, of which the export-processing industry was a main part, recorded an annual rate of 23 percent, significantly higher than the regional average of the Pearl River Delta (Guangdong, Statistical Bureau 1991c:238-41). The export-processing fees received by Dongguan increased from US$2.34 million in 1979 to US$163 million in 1990, an annual growth rate of 53.5 percent (CCP Team 1989:31; Guangdong, Statistical Bureau 1991b:357). Per capita income rose substantially from 193 yuan to 1,359 yuan for peasants, and from 547 to 3,552 yuan for salaried workers in the twelve years between 1978 and 1990 (Guangdong, Statistical Bureau 1991b:238-41). This extraordinary process of economic structural change and accelerated growth since the late 1970s has been unprecedented in Dongguan's history and was clearly fuelled by the inflow of investment and manufacturing facilities from Hong Kong and overseas.

An interesting phenomenon especially evident in Dongguan as a result of export industrial development is the increasing participation of women in manufacturing. Since export production is predominantly labour intensive in nature, its rapid expansion has opened much room for the employment of women who are generally considered nimbler than men for such jobs as toy making, apparel sewing, or electronic-products processing. A growing number of women have, therefore, joined this army of factory workers and are playing a part in the process of industrialization. In 1989, of the 166,000 workers employed by export-processing firms, 130,000 were women, accounting for 78 percent of the total work force (CCP Team 1989:159). In many workshops, workers are almost entirely female with only a few men responsible for repairing machinery, maintaining factory security, loading and unloading finished products or imported materials, and doing managerial work. Most female workers are under twenty-five. Some of them have begun to earn incomes equivalent to men's.

Women's participation in manufacturing production has undoubtedly raised their economic and social status, but it has also placed them in a confined environment where they are asked to work repetitively on the same piece at a desk for long hours and at a piecework wage. For those who are already married, factory work and household affairs have combined to form an almost unbearable burden. For those who are young, entering the factory at an early age means that they will have little chance of receiving further education and, therefore, few career or advancement alternatives. The intrusion of global market forces from Hong Kong has thus led Chinese women, who might otherwise have been housewives or college students, to take part in the process of a new international division of labour.

Another distinct demographic feature that has characterized the transformation of the local economy is the rapid growth of immigration, which is also a direct outcome of flourishing export-processing activities. By the mid-1980s, rapid expansion of the labour-intensive processing industry had exhausted the local supply of labour and created a large demand for outside workers. With the relaxation of government control on migration, which took effect in 1984, labour began to flow in from the less-developed counties of Guangdong and the interior provinces. Since the mid-1980s, immigration of outside labour has grown substantially at 43 percent per annum. By the end of 1990, 'outside labour' (*wailai laogong*) had reached 655,902 persons, which almost equalled the local labour force (CCP Team 1989:6). Considering that outside labourers have an employment rate of 98.39 percent, which is higher than that of local labourers (76.02 percent) (*Yatai Jingji Shibao* 1992, August 2), almost half of Dongguan's economy has apparently been run by hardworking outsiders. This fact is especially evident in the manufacturing sector, where 63 percent of the labour force was from outside. In some areas such as Chang'an zhen (township) at the border with Shenzhen, the number of migrants reached 91,000 in 1992, almost four times the local population.

By far, the vast majority of migrants to Dongguan have been engaged in manufacturing production, particularly in export processing. Statistical data have shown that about 80 percent of the total outside labour force, or 518,971 out of 655,902, was found in the manufacturing sector (Dongguan, Statistical Bureau 1991:6). A survey conducted in 1988 revealed that 61.12 percent of all outside labour was in export-processing plants (CCP Team 1989:95). Of all factory jobs created by the export-processing industry from 1979 to 1990, 85 percent was taken by migrants (*Yatai Jingji Shibao* 1992, August 2).

Many migrants to Dongguan are young women between eighteen and twenty-five, frequently referred to by the local people as 'working girls' (*da gong mei*) or 'girls from outside' (*wai lai mei*). They usually share dormitory rooms near the factory with eight to twelve persons, and they pay rent or a 'managerial fee' (*guanli fei*) to local cadres responsible for the construction and maintenance of both the factory and the dormitory buildings. The money they save is sent back via banks or postal offices to their relatives in poor interior areas. As a result, those townships that have a large number of outside workers tend to have a disproportionately large number of banks and post offices. In Chang'an zhen, where field investigation for this study was conducted, the main street of the town, a couple of hundred metres long, has fourteen banks open from 8 a.m. to 9 p.m. to serve outsiders who want to deposit or mail their savings to their hometowns. Some of these outsiders work in Dongguan for several years until they earn enough money to go home. Others stay for a prolonged time. A few have married local residents or set up their own businesses.

The experience of working and living in an environment surrounded by strangers is not, however, always pleasant for the outsiders. They frequently find themselves faced with discrimination, as the best jobs with higher pay always go first to locals and they get only the least desirable jobs. Speaking a completely different language from the local dialect, they can barely communicate with the local residents, and loneliness is something they have to get used to. As well, they can be cheated when they go shopping. Some of the 'working girls' from outside even have to bear what Westerners would call harassment or assaults from local factory managers or Hong Kong bosses who simply want to take advantage of them, or, in the words of the local people, bosses who 'treat the girls like a piece of pliable beancurd' (*chi ruan doufu*). The issue of 'working girls' from outside has become such a national concern that a number of movies and TV programs have been produced to show the unhappiness and bitterness of these newcomers. The most popular TV movie, which won the top national award in 1992, was named *The Working Girls* (*da gong mei*). The penetration of global market forces through Hong Kong into Dongguan has thus not only promoted the participation of local peasants and women in manufacturing production but also effectively drawn the young and cheap labour of China's interiors into the theatre of mass production and global capitalism.

The rapid growth of export production and its subsequent economic and demographic changes highlighted here are manifest over the geographical space. With a locational focus in the countryside, the development of the export-processing industry has inevitably resulted in a process of land-use transformation, in which much farmland has been turned over for the construction and expansion of factories. Data obtained from the Agricultural Department of Dongguan reveal that in 1978-88, a total of 18,585 mu or 3,061 acres of farmland was transformed for industrial land use, mostly for the building of export factories, workshops, and industrial districts (Huang 1991:79). Consequently, per capita cultivated land dropped substantially from 1.06 mu in 1978 to 0.67 mu in 1990.

Many small workshops and factories developed in the early 1980s were scattered over townships and villages in the countryside. As production expanded, local officials began to realize that such a spatial arrangement made it difficult to provide electricity, water, and sewage-disposal facilities. A new type of industrial land use has since gradually emerged, covering a sizeable amount of land and located at the outskirts of towns or villages along trunk roads. By the end of 1987, a total of 119 such industrial zones had emerged. In Chang'an zhen, for example, one such industrial zone covers 198 mu or 32.6 acres. Developed jointly by Chang'an zhen and several Hong Kong companies, this zone absorbed a total investment of 236.50 million Hong Kong dollars or US$30.32 million and accommodated over 1,000 employees working on the spinning, weaving, and dying of textile materials for Hong Kong manufacturers.

Such newly emerged industrial zones are usually built on a piece of farmland and are typically composed of a group of identical factory buildings of two or three stories. Each factory building belongs to a certain processing firm, and the name of that firm, in large red Hong Kong-style Chinese characters, can be seen at the top of the building. At the entrance of each factory stands a security guard who dresses in an uniform resembling that of a Hong Kong policeman. All factory buildings in the zone are arranged in straight rows and face the same direction. Infrastructure facilities for these factories are generally well established and are used more efficiently than those of individual factories that are built separately.

Most of these export-processing factories, built separately or in groups, are located in the vicinity of the headquarters of former communes and brigades. Their development and continuous expansion have greatly fostered the industrialization of land in the countryside and created a distinctive type of land use characterized by a mix of farmland, factories, and housing for peasants.

Such a process of industrialization did not, however, force those farmers who lost their land to move into the city. Instead, by creating factory jobs in the countryside, the growth of the processing industry has allowed peasants to 'enter the factory but not the city' or 'leave the soil but not the village.' Between 1978 and 1987, for instance, an estimated 154,000 persons, most of them surplus rural labourers, joined the industrial labour force of Dongguan. Among these new workers, only 34,000 or 22 percent went into factories in towns. The other 120,000 new workers or 78 percent entered factories and workshops in the countryside (CCP Team 1989:27). Clearly, most of those farmers who were released from traditional agricultural production obtained factory jobs near their villages without moving into cities and towns. Export industrial development, with its locational focus in the countryside, has undoubtedly contributed to shaping this spatial pattern, in which the transformation of land and population in the countryside has been no less significant, if not greater, than that in the city. It also provides a good explanation for the distinctive spatial features of development in the Pearl River Delta region, outlined in Chapter 5.

The Social Influence of Hong Kong

The influence of Hong Kong has gone beyond the economic sphere and provoked important social changes in the townships and villages of the Pearl River Delta. Significant changes in the lifestyle in Guangdong Province and particularly in the Pearl River Delta as a result of the influence of Hong Kong have been documented by scholars such as Vogel (1989) and Guldin (1992). This section provides a local example to reinforce Vogel and Guldin's early arguments.

With its frontier location and excellent connections to Hong Kong, Dongguan is one of the first among the cities and counties of the delta that has felt the strong 'south wind' from Hong Kong, which brings the air of capitalism into socialist territory. As the special identity of Dongguan, the development of an export-processing industry since 1979 has ushered in new production systems, new management styles, and a new job attitude. In the processing plants that subcontract from Hong Kong manufacturers, workers have no promise of job security, any rewards are tied to the amount of work finished, the time requirements are rigid, and the pressure on workers to keep up a fast working pace is high. For those used to the socialist production system under which job security or 'iron bowl' is guaranteed, work in such a subcontracting factory or joint venture means fundamental changes in job-related attitudes, values, and behaviours. Doing a factory job is no longer considered as fulfilling the glorious socialist obligation of 'serving the people,' but instead is a way of earning a living for personal gain. Because nothing can be counted on, people have become more independent, efficient, and sensitive to changes in their living environment. At the same time, loneliness, frustration, and depression over not being able to keep up with the working pace or to realize personal ambition have become increasingly noticeable in the local community.

Visitors from Hong Kong who went to Dongguan to do business or see relatives often brought with them ideas, information, and different lifestyles. Since its opening up in 1979, Dongguan has been visited more frequently than ever before by relatives from Hong Kong. In 1990, for instance, 262,586 visitors entered Dongguan from Hong Kong, for business or family affairs (Guangdong, Statistical Bureau 1991c:271). These visitors always brought information and materials to their Dongguan relatives, allowing them to share in the Hong Kong consumerist vision of modernity. In the early 1980s, when modern consumer goods such as TVs, VCRs, and refrigerators were still rarely seen elsewhere in the nation, Dongguan had already started to receive a variety of gifts from Hong Kong kinfolk, including washing machines, TV sets, VCRs, hi-fi stereos, motorcycles, and fashionable clothes. Over the years, Dongguan has received so many consumer goods from Hong Kong that its population owns more colour TVs, stereo tapedecks, and washing machines than people in other cities and counties of the province. A 1990 survey conducted by the provincial authorities revealed that town residents in Dongguan owned 112 colour TV sets, 102 stereo tapedecks, and 94 washing machines per 100 households, higher rates than all of the surveyed households in other cities in the province, including Guangzhou, Foshan, Shenzhen, and Zhuhai (Guangdong, Statistical Bureau 1991a:370-1). As well, the ownership rates of colour TVs, motorcycles, refrigerators, and stereo tapedecks for the peasant households of Dongguan in 1990 were also significantly higher than the provincial averages (Guangdong, Statistical Bureau 1991a:386). Needless to say, these consumer goods have provided a material basis for imitating the Hong Kong lifestyle in Dongguan. Modern electronic receivers such as TVs, radios, and VCRs are also important conduits for the penetration of Hong Kong culture into the towns and villages of Dongguan.

The Hong Kong lifestyle is probably most effectively conveyed to the local people by the modern mass communication network that links Hong Kong to almost all Dongguan households. The computerized telephone system, which was installed in 1984 and has rapidly expanded ever since, allows Dongguan's residents to dial direct to their relatives in Hong Kong and overseas for information about the outside world. Electronic conduits such as TV and radio have brought almost all programs broadcast from Hong Kong stations into nearly all Dongguan households. For the first time since the founding of the People's Republic, Dongguan's residents are able to receive sounds and images about the life of their relatives on the other side of the border, a lifestyle in sharp contrast to what they have been used to for over thirty years.

When my field research was first conducted in 1984, local households with TV sets could watch the Hong Kong program of *Foon Luk Gum Siu* (*Enjoy Yourself Tonight*), an imitation of the American *Tonight Show* that starred Johnny Carson. Few of Dongguan's residents had any interest in tuning into the Central Broadcasting Station of Beijing for Communist propaganda or government-controlled news. When I revisited Dongguan in 1992, I was stunned by the fact that many peasants in rural villages were able to watch all sorts of American TV programs broadcast from the Pearl Station of Hong Kong, including the *CBS Evening News*, *60 Minutes*, *20/20*, *Wall Street Journal*, *Dallas*, *Murphy Brown*, and *America's Funniest People*, as well as many American movies in bilingual form (English/Cantonese).

The invasion of Western culture, ideas, and information from Hong Kong into Dongguan has begun to alter the existing landscape and change the lifestyle of the local people. Visitors to Dongguan who travel around the countryside can easily find a distinctive landscape of numerous large antennas, erected at the top of farmhouses to capture coveted TV signals from Hong Kong. Within a village or town, the central location is usually occupied by a 'cultural centre' (*wenhua zhongxin*), which was originally set up by the former commune or brigade officials to popularize Marxism, Leninism, and Mao Zedong Thought, or to disseminate directives from Beijing. Ironically, these socialist 'cultural centres' have surrendered to Hong Kong's influence and become the locus for popularizing recreational activities in the Hong Kong style. Instead of studying the quotations of Chairman Mao or reading the editorial comment of the *People's Daily*, people come to the cultural centre to play video games, disco dance, sing Hong Kong pop songs Karaoke-style, and see video movies smuggled in from Hong Kong. Action and horror movies, and even adult movies, are no longer uncommon, and most of them are produced in Hong Kong.

From TV, radio, and other media, the local people of Dongguan have become increasingly familiar with product brand names and have begun to consume foreign goods such as Colgate toothpaste, Marlboro cigarettes, Nike sneakers, and foreign-made cosmetics. Foreign-made drinks such as Pepsi-Cola, Coca-Cola, 7-Up, and Maxi coffee, which had never been heard of before 1979, have become the most common items for daily consumption by the local population.

Another interesting reflection of Hong Kong's influence can be found in changes in the local language characterized by the frequent use of many words from a Hong Kong version of English. Thus, 'bye-bye' has replaced the traditional Chinese saying of *manzou* (please walk slowly). Instead of calling policemen *min jing* (the people's guard), local people have begun to adopt the Hong Kong Cantonese English 'loan-words' of *ah Sir* (Sir) or *chai lo* (servant fellow) to refer to the police. Other Hong Kong translations of English such as *desi* (taxi), *shido* (store), and *salong* (saloon) have also become popular. On entering a restaurant in Chang'an zhen of Dongguan, I was puzzled by many dish names on the Chinese menu that I could not decipher, as they were neither in English nor in Mandarin. They were all instead in a Hong Kong sound translation of English such as *sari* (salad), *buding* (pudding), *pisabing* (pizza), and *bingqiling* (ice cream). What has been taking place in Dongguan since its opening up in 1979 is the demise of socialist tradition, the weakening of the impact of Beijing both politically and culturally, and the strengthening of the Hong Kong model of production, consumption, recreation, and communication. Given sufficient time, a unique culture that blends the local tradition with Hong Kong innovations in dress, speech, music, and lifestyle may well emerge in Dongguan as well as in other parts of the delta region.

Summary

For three decades after the birth of the People's Republic, socialist China adopted an inward-looking approach to national and regional development, primarily due to the existence of a hostile international environment. With its frontier location and perceived vulnerability to overseas attacks, the Pearl River Delta region had not been able to receive adequate support from the state for industrial development or to perform its traditional role as a centre of export production and international trade. The implementation of the open door policy in 1979 renewed and strengthened the delta's economic and social linkages with Hong Kong and overseas, which resulted in fundamental changes not only in the economic sphere but also in spatial and cultural aspects. If anything distinguishes the recent development of the Pearl River Delta from other parts of the nation, it is the rapid and widespread penetration of global market forces from its capitalist neighbour, Hong Kong. This should be carefully studied in order to understand economic and spatial changes occurring in this region.

This chapter assesses the role played by external forces in the process of the economic restructuring and spatial transformation of the Pearl River Delta, using Dongguan as a case study. My detailed investigation has revealed that the relocation of labour-intensive industrial production from Hong Kong to Dongguan has created considerable employment opportunities in manufacturing for the local population, most of whom were farmers, and, therefore, significantly facilitated the transformation of the local economy from predominantly agricultural into greater reliance on manufacturing and service sectors.

Manufacturing facilities relocated from Hong Kong have not been concentrated in a few urban centres, as the prevailing theory of multinational corporations would suggest. The existence of personal kinship ties, an improved transport and electric power infrastructure, an abundant supply of cheap labour and land space, and the lack of strict regulations on environmental pollution have all made the countryside a place no less attractive than a congested large city to multinationals from Hong Kong and overseas. In Dongguan, a vast majority of the small-scale, labour-intensive, and technologically unsophisticated processing plants subcontracting for Hong Kong companies has been found widely spread over the rural townships and villages of the countryside. Such externally driven and rural-based industrial growth has resulted in a process of spatial transformation, where a great number of surplus rural labourers have entered factories in the countryside without moving into the city, and much agricultural land has been converted into factory sites or industrial zones. The rapid expansion of the export-processing industry has also contributed to accelerated growth in production, a substantial increase in personal income, a growing participation of women in manufacturing, and a massive inflow of migrants from China's interior, all of which are no less significant than what has occurred in such large cities as Guangzhou, Foshan, and Jiangmen. The penetration of economic forces from Hong Kong as a result of both China's open door policy and global industrial restructuring is one of the contributing factors that explains why, as identified in Chapter 5, the magnitude of growth and spatial change has been remarkably greater in the suburban counties of the Pearl River Delta region than in the large cities.

The opening up of the Pearl River Delta, originally intended to attract foreign investment and acquire advanced technological know-how, has also exposed the traditional culture of the local community to the filtering in of Hong Kong-style consumption, recreation, and communication. The influence of Hong Kong and the subsequent social and cultural changes discussed in this chapter are not confined to Dongguan alone. They can also be found in other cities and counties of the delta where extensive kinship, linguistic, and other cultural ties with Hong Kong have long existed. Social and cultural connections between the delta region and Hong Kong will continue to reinforce each other with economic linkages and combine to shape the emerging Hong Kong-Guangzhou megalopolis.

9
Conclusion

Against the backdrop of a rapidly collapsing socialist empire in Eastern Europe and the former Soviet Union, China under the pragmatic leadership of Deng Xiaoping has since the late 1970s endeavoured to develop a 'socialist market economy with Chinese characteristics.' If the 1980s can be recorded as a turning point in the history of global development, because the decade marked the dissolution of the long-lasting socialist bloc and the end of the Cold War, it should also be considered a landmark in contemporary Chinese history, as it featured the abandonment of a Maoist interpretation of socialism, the liberation of the socialist economy after thirty years of rigid state control, and an unprecedented outcry for political freedom, democracy, and human rights. Indeed, recent developments in China since the 1980s have been so profound and dramatic that they deserve systematic assessment and scholarly scrutiny. Such developments also constitute a rare and valuable case to show how an isolated socialist economy is gradually transformed rather than abruptly destroyed after the intrusion of global capitalist forces.

This study of economic and spatial transformation in China since the reforms has focused on the case of the Pearl River Delta in South China. Systematic analyses of regional data and detailed case studies have revealed that since 1978 the delta region has undergone a process of dramatic structural and spatial transformation. Its proximity to Hong Kong, abundant agricultural resources, and decentralized economic decision-making have all combined to enable the delta to move 'one step ahead' of the nation to attract foreign capital investment, achieve incredibly rapid growth in industrial and agricultural production, and significantly raise productivity, employment, and income for its population. After decades of isolation, stagnation, and 'involutionary growth' or 'growth without development,' the Pearl River Delta has, for the first time since the People's Republic was established, displayed remarkable signs of genuine transformative development.

The rapid development of the delta regional economy has owed little to the expansion of state-run, large-scale, capital-intensive modern manufacturing. Instead, numerous rural-based, small-scale, and labour-intensive industries financed by both local and foreign capital have provided the primary energy to fuel the process of industrialization and regional development. The structural consequence of this rural-based industrialization has been the disproportionate growth of the manufacturing sector and the simultaneous decline of agriculture in terms of total output. The agricultural sector was completely wiped out after industrialization. Instead, it has been able to commercialize and diversify to meet new challenges and become a profitable economic sector. The persistence of agricultural production in the Pearl River Delta is primarily due to the unique local conditions, including excellent natural endowments, a well-established farming tradition, and easy access to major urban markets overseas. A restructured dual economy featuring both rural industry and commercialized agriculture is gradually taking shape.

The spatial configuration of economic development in the delta has been region-based rather than city-based urbanization. This region-based urbanization has displayed distinct features in two simultaneous interrelated processes: population redistribution and land-use transformation. In terms of population redistribution, rapid growth and restructuring of the regional economy did not result in a growing concentration of population in large cities, as the conventional wisdom of urban transition might have predicted. The primate city of Guangzhou did not exhibit any accelerated growth in production or population. It is the triangle bordered by Guangzhou, Hong Kong, and Macao that has attracted a growing population. Within the regional hierarchy of urban settlements, it is the numerous small towns that have absorbed a great majority of the surplus rural labour released from agricultural production.

Similarly, the process of land-use transformation is better attributed to the industrialization of the countryside than to the expansion of large cities. A substantial amount of farmland has been lost to industrial, transport, residential, and commercial developments in the countryside. A direct spatial outcome of this region-based rather than city-based urbanization has been mixed industrial/agricultural land-use in the countryside and a blurring of the distinction between urban and rural activity.

My detailed case studies of the operating mechanism of spatial transformation suggest that economic growth and structural changes in this southern China region have not been the outcome of any active state involvement in local economic affairs. Rather, they have been the result of relaxed state control over the local economy. Economically, the central state no longer monopolizes production, circulation, and distribution (*shengchan, liutong, fenpei*). Financially, the central state has contributed only a very small proportion of the capital for local infrastructure development. The bulk of capital has actually been mobilized by county and township governments through bank loans and overseas channels. Geographically, the people of the Pearl River Delta may have more flexibility than those elsewhere in the country in deciding what and how things should be done, given that 'the mountains are high and the emperor is far away.' Culturally, the delta region has since its opening up become more subordinate to Hong Kong than to Beijing. If the central state has contributed anything, it is the tacit laissez-faire approach to regional issues that has allowed the Pearl River Delta to experience virtually self-motivated, self-sustained, and market-oriented genuine development.

While the role of the central state has been significantly weakened as a result of economic reforms, local initiative has begun to take over and steer the vehicle of regional development. In particular, county and township governments as well as the collective and private sectors have become the chief agents responsible for two remarkable processes: rural industrialization and transport development. It is the local governments at the county, township, and village levels that have mobilized necessary capital, labour, and land resources to develop rural industries and market farming. These changes have enabled the region to break the deadlock of involutionary growth and enter a new stage of transformative development. County and township governments are also the key players in the creation of a transactional environment conducive to regional development. By building numerous bridges, roads, and highways, local economic planners have managed to overcome the 'friction of distance' and rearrange economic activities over space in a way deemed necessary and instrumental to the pursuit of their development ambitions.

The influence of global capitalism emanating primarily from Hong Kong is perhaps the most distinctive feature of regional development in the Pearl River Delta. Such influence is manifest in the two spheres of economic transformation and social change. Economically, the establishment of joint ventures, compensation trading firms, and other forms of economic cooperation has significantly facilitated the transformation of local labour, induced immigration, and increased the pace of farmland loss. Socially, the filtering of information, ideas, and modern technology from Hong Kong into the delta has Westernized the lifestyle of the local people, particularly in their consumption, recreation, and communication. The external forces of global capitalism have not gone head to head with the local economy or society. Instead, they have been grounded in some local conditions, particularly personal networks, for reliable economic returns. This distinctive manner of local-global interaction underlies the process of the socialist economy transforming gradually rather than abruptly in response to capitalist forces. It also gives rise to a spatial pattern in which incipient capitalism first takes root in a few selected areas (not necessarily in large cities) where local conditions are favourable to its growth.

Theoretical and Planning Implications

Until recently, various theoretical interpretations of the dynamics of regional development have been by and large polarized into two schools of thought, namely exogenism and endogenism. In the early neo-classic period of polarized economic growth, regional development was seen as motivated by external demand and driven by innovation impulses. The conviction was that regional development only starts in a few dynamic sectors and geographical clusters. Once started, the benefits of polarized growth will, in a spontaneous or an induced way, 'trickle down' to the rest of the spatial systems, leading to an eventual convergence of regional disparities. The role played by external forces such as demand, investment, or innovation was thus seen as essential, positive, and generative.

The persistence of poverty and growing income inequality after decades of development resulted in disillusionment with the optimistic speculation about the convergence of social and spatial disparities. The dominance of the dependency school in the late 1960s meant a reversal in the evaluation of external forces, from the previous positive or generative assessment to a negative or parasitic formulation. However, the underlying exogenous assumption remains unchanged. The pattern of development and underdevelopment was still perceived as being shaped by external forces.

Theoretical advances in the post-dependency era have shown significant signs of reorientation from the previous bias of exogenism to endogenism. This process of reorientation, called 'indigenization' by Hettne (1990:243), is marked by a shift of focus from external forces to intra-regional or domestic conditions. This trend of reorientation is evident in the notions of 'another development,' basic need satisfaction, and agropolitan development, which all stress the importance of internal mobilization of local indigenous resources. The world-system thesis, which essentially internalizes the external factors by changing the scale of analysis, has also been considered 'a return to endogenism of a more grand scale, the endogenism of the world-system rather than the endogenism of the nation state' (Hettne 1990:245). The neo-Marxist doctrine, although analyzing within a broader global and historical context, has shown an apparent orientation toward endogenism (Cowen and Shenton 1996:140; Hettne 1990:244; Corbridge 1989:231). In explaining the pattern of development and underdevelopment, for instance, Marxists have maintained that new and higher production relations could not appear 'before the material conditions of their existence have matured in the womb of the old society' (quoted by Hettne 1990:244). Development thus perceived is virtually an internally driven, automatic, and repetitive process, in which the underdeveloped countries fulfil the necessary 'material conditions' so as to reach a higher production relation.

The experience of the Pearl River Delta region in the past decade suggests that neither exogenism nor endogenism can provide a satisfactory explanation for the dynamism of regional development. Instead, it is the dialectical interaction between local and global forces that has created the complex scenario of changes in the Chinese spatial economy. Figure 9.1 summarizes the main features of regional economic transformation. This model has three interconnected spheres. The first sphere concerns the interaction between local and global forces. It starts with the relaxed control of the state over the regional economy, which has enabled both local initiative and global market forces to shape the spatial pattern of growth and development. Local initiative is the main driving force responsible for transforming the rural economy and setting up an economic infrastructure. On the other hand, the penetration of global capitalist forces is manifest in the growing influence of Hong Kong on the development of the delta. These three processes of rural industrialization, transport development, and Hong Kong's influence have acted together in the restructuring of the spatial economy of the delta region. Rural industrialization and agricultural marketization would not have been possible without a developed transport infrastructure and the urban mar-

Figure 9.1

A model for spatial transformation in the post-reform Pearl River Delta, 1980-90

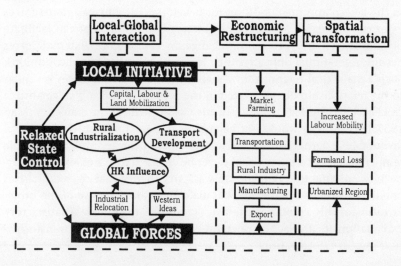

ket of Hong Kong, which provides a great demand for the delta's manufactured products and farm commodities. Similarly, transport infrastructure development owes a great deal to Hong Kong, which provides a considerable amount of capital, technology, and management know-how to
the delta. Reciprocally, the expansion of the transport infrastructure and
market-oriented rural industrialization reinforces the influence of Hong
Kong on the delta region. Thus, the three essential processes of local-
global dialectics are interdependent.

The consequence of interaction between local and global forces has
been the rapid growth and restructuring of the regional economy and the
spatial redistribution of population and land use, which are identified as
the second and third spheres of the model. Economically, the transformation of the rural economy has significantly promoted both rural industrial
production and market farming at the expense of traditional paddy rice
cultivation. The expansion of the transportation infrastructure has, needless to say, raised the importance of the transport sector in employment
and revenue generation. At the same time, foreign investment, mostly
from Hong Kong, has contributed to the rapid growth of manufacturing
and export production. The spatial consequence of the interaction of local
and global forces has been faster population movement to and within the
region, the transformation of land from farm to non-farm uses, and the
creation of a urbanized zone stretching along the Hong Kong-Guangzhou-
Macao corridor (see Figure 9.1).

From a theoretical perspective, the model of local-global interaction
derived from the experience of the Pearl River Delta suggests a dialectical view of the operating mechanism of regional development as an
alternative to the existing theories of exogenism and endogenism. In
this model, local initiative, or the internal force of change, is perceived
as the key to and pre-condition for development, whereas global forces
external to a region are seen as secondary conditions that can facilitate
or slow down the process of change in the region. External forces
become operational only after their interaction with internal conditions.
Such a dialectical relationship between local and global forces is similar
to the relationship demonstrated in the process of turning an egg into a
chicken cited by Mao Zedong in his classic writing on dialectics (Mao
1937:157). An egg would not be able to change into a chicken unless a
favourable external condition was provided (for example, a suitable
temperature), but a stone could never be changed into a chicken no matter how suitable the temperature was. Any theory that overemphasizes
one dimension of the local-global dialectic at the expense of the other,
or mechanically combines the two, would fail to provide a satisfactory
explanation for the actual operating mechanism of spatial transformation in post-reform China.

This study of the economic and spatial transformation of the Pearl River Delta region has also raised some fundamental questions about the role of the socialist state in national and regional development. As I have discussed elsewhere (Lin 1994a:15), existing literature on the issue of socialist regional development has basically centred on the two opposing assertions of egalitarianism and pragmatism. The former perceives the socialist pattern of limited urban and regional growth as a product of the Marxist ideological commitment to anti-urbanism and spatial equality, whereas the latter stresses some practical factors such as urban manageability and national defence (Kirkby 1985; Cannon 1990; Chan 1994). Common to both assertions is the fundamental assumption that the socialist nation state has always played a pivotal role in national and regional development, although the motives of the state are interpreted by some as supporting egalitarianism or anti-urbanism, but understood by others as seeking economic rationality and national security. The development experience of the Pearl River Delta clearly suggests that the prevailing assumption of a strong socialist state may no longer be valid for what has been taking place in post-reform China, although it might hold true for urban and regional growth in the Maoist era. Due to the 1978 economic reforms, which decentralized economic decision-making from the central state to local governments and individuals, the role of the central state in regional economic development has been significantly weakened. In view of the changing role of the central state, many theoretical generalizations on socialist development must be reassessed and updated. We need to know, for instance, if weakened state intervention is a feature unique to South China or common to other Chinese regions where a market economy is quickly developing. We also need to know the spatial and social consequences of economic decentralization at a national scale. Studies of these issues will enrich our knowledge of the interplay between political economy and regional development, and enhance our understanding of the dynamics of spatial transformation in the context of reformed socialism.

From the perspective of regional planning and management, the tacit laissez faire policy adopted by the post-reform socialist state has not always been a blessing to the development of the national and regional economy. While relaxed state control has undoubtedly stimulated local incentives and enthusiasm, it has also resulted in many problems that local authorities are unable to handle. Many small rural industries, for example, have in recent years found themselves 'swimming in the billowing ocean of the commodity economy' without any idea of where they are heading. These small industries are able to respond sensibly to the changing demands of the market, but they are unable to detect and monitor changes in the larger economic environment of which they are a part. As a result, after a certain commodity is perceived as in great demand, numerous enterprises act like 'a swarm of bees' (*yiwofeng*) to produce that commodity. The market is then quickly saturated and many producers soon suffer from excessive stock, sluggish sales, and, eventually, bankruptcy.

Relaxed state intervention has also allowed peasants to transform most of their fertile cultivated land into more profitable industrial, commercial, and real estate developments. Profits generated from non-agricultural activities are then used to purchase food grain from other provinces or overseas to satisfy local demand. This smart arrangement seems to be working well for now, but in the long run the local people of the Pearl River Delta will have to pay a price for uncontrolled and irreversible farmland loss. Real estate development, which has taken a tremendous amount of farmland, is already a hot business in many cities and counties of the delta. Once it cools down, it will not be able to generate enough profit to sustain the agricultural imports local people now expect.

Loosened state intervention has opened the door for such problems as environmental pollution, chaotic land use, social insecurity, burglary, smuggling, prostitution, counterfeiting of money, and the illegal simulation of brand-name products including drugs and medicines – all of which have devastating effects on the lives of local people. Without the necessary macro-level control, including legislative, economic, and financial regulations, the Pearl River Delta will be unable to achieve economically viable, socially stable, and environmentally sustainable long-term regional development.

The detailed case study of transport development in the delta region, particularly in Panyu, also shows an interesting pattern that has significant theoretical implications. For decades, the nature of the relationship between transport expansion and economic growth has been a subject of prolonged inquiry and unresolved debate. It has been generally recognized that there are three different interpretations of the role of transport expansion in regional development. The first school of thought takes a neoclassic economic stance and sees transportation as a prerequisite for the expansion of a regional economy (Owen 1964, 1987; Taaffe and Gauthier 1973). The second vision subscribes to the dependency school and perceives transport expansion as having a negative effect on regional growth, because this expansion could absorb a large amount of scarce capital that should be invested elsewhere to satisfy the pressing basic needs of the general population (Wilson 1966; Gauthier 1970). The third interpretation takes a permissive perspective and contends that transportation has no causal effect on economic development (Leinbach and Chia 1989). The experience of the Pearl River Delta region has suggested that the establishment of a transport infrastructure in the right economic environment can have a remarkable effect, leading to rather than following the transformation of a local economy. The effect of transport expansion has been especially evident in Panyu, where economic growth had long been blocked by the lack of a modern transport infrastructure. The frontier extension of road transportation has created definitive effects on the process of spatial transformation, including accelerated suburbanization of production, population, and human settlements. The experience of the Pearl River Delta region, where economic growth is still at an early stage, has been fundamentally distinct from that of regions in Western Europe and North America, where development has already reached an advanced stage of maturity and where transport improvement is usually seen as a response to further economic expansion. This finding suggests that an appropriate, or perhaps the only, way of understanding the complexity of interaction between transport development and regional economic transformation is to place the issue in a historically and geographically specific context for assessment. As well, it implies that to assume a definite cause-effect relationship applicable in all cases would be an oversimplification.

Finally, my analysis of the flow of foreign capital and manufacturing facilities, mostly from Hong Kong, into the Pearl River Delta has revealed that the dynamism of transnational capital has been actually more complicated than what has been suggested by the conventional wisdom of globalization. Thus far, conceptualization of the global shift of capital investment and manufacturing production has been primarily economic and oriented toward large cities. Transnational movement of capital is usually understood as driven by such economic incentives as reducing labour costs, increasing net profits, and strengthening competitiveness. Geographically, transnational capital is described as having a locational preference for large cities, where a good infrastructure is readily available and where economic externalities prevail. As a result, transnational capital is believed to have a direct spatial effect on the accelerated expansion of large cities in the target region. This theory of globalization may provide a general account for the pattern of investment at the global scale, but it may not be a sensible and adequate explanation for the detailed pattern of spatial distribution at a regional or local scale.

In the delta region, the growth and the spatial distribution of capital investment and manufacturing production relocated from Hong Kong have not been completely shaped by economic forces. Many manufacturing establishments that have subcontracted with Hong Kong firms have developed primarily on the basis of personal connections or kinship ties that have enabled investors to secure a reliable return on their investment (Leung 1993 1996; Smart 1993; Smart and Smart 1992). Since the existence of kinship ties has not been limited to large cities, and since most relocated manufacturing operations are simple, small, labour intensive and geographically ubiquitous, the transnational capital and manufacturing production subcontracted with Hong Kong have not displayed a strong tendency to concentrate in large cities. They have actually favoured the suburban corridor between Hong Kong and Guangzhou, and have become a significant contributing factor in explaining why the large cities of the Pearl River Delta have not experienced excessive expansion since the reforms, whereas many suburban counties have been growing rapidly. This finding suggests that the transnational movement of capital and manufacturing is not a purely economic phenomenon and that non-economic factors such as historical, cultural, and social relations should not be overlooked in understanding the mechanism and spatial patterns of the globalization of production.

Prospects

The economic and spatial changes that have been taking place in the Pearl River Delta since the 1980s are truly staggering and fascinating. As for this 'land of capitalist miracles' – where rice paddy is being transformed into metropolis, where thousands of factories are blooming, where roads are being extended so rapidly that no map is accurate, where hotels and restaurants are being opened every day, and where telephone directories and company listings are immediately dated – no one can be sure what its future will hold.

In the first decade of reforms, the Pearl River Delta was chosen as one of the first regions in the nation to practise capitalism and develop an open market economy. With the recent opening up of other coastal regions in the country, such as Pudong in Shanghai, the Pearl River Delta region will gradually lose its leading position in attracting foreign investment and developing a market economy. Its access to national and international markets will also be constrained, because other Chinese coastal regions and other Southeast Asian countries are developing rapidly and offering increasing competition. The first decade of reforms has pulled the Pearl River Delta out of the deadlock of isolated and stagnant growth. The challenging task confronting the planners of the delta is to develop a strong regional economy in order to win in both national and international competition. Effort has to be made to resolve internal economic problems, such as chaotic production, inflation, and irrational real estate development, to consolidate the foundation of the region's economy according to its comparative advantages, and to sharpen up several key industries that have regional strengths.

It may not be possible to accurately predict the future of the Pearl River Delta, because the region is, like the nation as a whole, undergoing another dramatic transition after the return of Hong Kong to Chinese sovereignty. It is foreseeable, however, that the Pearl River Delta, with its development headstart over other Chinese regions, will continue to serve as a rare laboratory for testing the validity of many theoretical hypotheses about development and for assessing the effectiveness of different policies or planning approaches. From a geographic standpoint, it would be of great interest to see if the development experience of the Pearl River Delta will also appear in other less-developed regions of the country after their economies have been exposed to similar global capitalist forces from Hong Kong and the Western world. Given China's regional diversity, the role of local forces in economic and spatial development may also vary. If so, we must consider whether one general national model, or several specific regional models, are required to account for the realities of economic and spatial transformation in post-reform China.

References

Publications in English

Abler, R. 1975. 'Effects of space-adjusting technologies on the human geography of the future.' In *Human Geography in a Shrinking World*, ed. R. Abler, D. Janelle, A. Philbrick, and J. Sommer, 35-56. North Scituate, MA: Duxbury Press

Aglietta, M. 1976. *A Theory of Capitalist Regulation*. London: New Left Books

Armstrong, W., and T.G. McGee. 1985. *Theatres of Accumulation: Studies in Asian and Latin American Urbanization*. New York: Methuen

Ash, R.F. 1988. 'The evolution of agricultural policy.' *China Quarterly* 116:529-55

Asia Research Centre, Murdoch University. 1992. *Southern China in Transition*. Canberra: East Asia Analytical Unit, Department of Foreign Affairs and Trade

Baster, N., ed. 1972. *Measuring Development*. London: Frank Cass

Blomstrom, M., and B. Hettne. 1984. *Development Theory in Transition*. London: Zed Books

Boudeville, J.R. 1966. *Problems of Regional Economic Planning*. Edinburgh: Edinburgh University Press

Bourne, L.S. 1991. 'Recycling urban systems and metropolitan areas: A geographical agenda for the 1990s and beyond.' *Economic Geography* 67 (3):185-209

Brookfield, H. 1975. *Interdependent Development*. London: Methuen

Brown, L.A. 1991. *Place, Migration and Development in the Third World*. London: Routledge

Brunn, S.D., and T.R. Leinbach, eds. 1991. *Coliapsing Space and Time*. London: Harper Collins Academic

Byrd, W.A., and Q.S. Lin, eds. 1990. *China's Rural Industry*. Washington, DC: World Bank

Cannon, T. 1990. 'Regions: Spatial inequality and regional policy.' In *Geography of Contemporary China*, ed. T. Cannon and A. Jenkins, 28-60. London: Routledge

Chan, K.W. 1992. 'Economic growth strategy and urbanization policies in China, 1949-1982.' *International Journal of Urban and Regional Research* 16 (2):275-305

-. 1994. *Cities with Invisible Walls*. Hong Kong: Oxford University Press

-, and X.Q. Xu. 1985. 'Urban population growth and urbanization in China since 1949: Reconstructing a baseline.' *China Quarterly* 104:583-613

Chang, S.D. 1981. 'Modernization and China's urban development.' *Annals of the Association of American Geographers* 71 (2):202-19

Chen, X. 1994. 'The new spatial division of labour and commodity chain in the Greater South China economic region.' In *Commodity Chains and Global Capitalism*, ed. G. Gereffi and M. Korzeniewicz, 165-86. Westport, CT: Greenwood Press

Cheung, P.T. 1994. 'The case of Guangdong in central-provincial relations.' In *Changing Central-Local Relations in China*, ed. H. Jia and Z. Lin, 207-38. Boulder, CO: Westview Press

Chu, D.K.Y. 1996. 'The Hong Kong-Zhujiang Delta and the world city system.' In *Emerging World Cities in Pacific Asia*, ed. F.C. Lo and Y.M. Yeung, 465-97. Tokyo: United Nations University Press

Cohen, R. 1981. 'The new international division of labor, multinational corporations and urban hierarchy.' In *Urbanization and Urban Planning in the Capitalist Society*, ed. M. Dear and A.J. Scott, 287-318. London: Methuen

Cohen, S.B. 1991. 'Global geopolitical change in the post-cold war era. Presidential address.' *Annals of the Association of American Geographers* 81 (4):551-80

Corbridge, S. 1989. 'Marxism, post-Marxism, and the geography of development.' In *New Models in Geography*, ed. R. Peet and N. Thrift, 224-54. London: Unwin Hyman

–. 1991. 'Third World development.' *Progress in Human Geography* 15 (3):311-21

Cowen, M.P., and R.W. Shenton. 1996. *Doctrines of Development*. New York: Routledge

Crook, F.W. 1986. 'The reform of the commune system and the rise of the township-collective-household system.' In *China's Economy Looks toward the Year 2000*, vol. 1, ed. Joint Economic Committee, Congress of the United States, 354-72. Washington, DC: US Government Printing Office

Dahrendorf, R. 1968. 'Market and plan: Two types of rationality.' In *Essays in the Theory of Society*, ed. R. Dahrendorf, 215-31. London: Routledge & Kegan Paul

Darwent, D.F. 1969. 'Growth poles and growth centers in regional planning: A review.' *Environment and Planning* 1:5-32

Dewar, D., A. Todes, and V. Watson. 1986. *Regional Development and Settlement Policy*. London: Allen & Unwin

Dicken, P. 1994. 'Roepke lecture in economic geography. Global-local tensions: Firms and states in the global space-economy.' *Economic Geography* 70 (2):101-28

Dwyer, D., ed. 1972. *The City as a Centre of Change in Asia*. Hong Kong: Hong Kong University Press

–. 1994. *China: The Next Decades*. Essex: Longman

Eckstein, A. 1977. *China's Economic Revolution*. Cambridge: Cambridge University Press

Edgington, D.W. 1986. 'China's open door policy: The Tianjin case.' *Pacific Viewpoint* 27 (2):99-119

Fan, C.C. 1995. 'Of belts and ladders: State policy and uneven regional development in post-Mao China.' *Annals of the Association of American Geographers* 85 (3):421-49

Fei, H.T. 1986. *Small Towns in China: Function, Problems and Prospects*. Beijing: New World Press

Foggin, P.M., C. Wang, and Z. Hu. 1993. 'Land use patterns in some medium-sized cities of the PRC.' Paper presented at the bi-annual conference of the East Asia Council, Canadian Asian Studies Association, University of Montreal

Forbes, D., and N. Thrift, eds. 1987. *The Socialist Third World*. Oxford: Basil Blackwell

Friedmann, J. 1972. 'A general theory of polarized development.' In *Growth Centres in Regional Economic Development*, ed. N.M. Hansen, 82-107. New York: Free Press

–. 1992. *Empowerment: The Politics of Alternative Development*. Oxford: Basil Blackwell

Frobel, F. J. Heinrich, and O. Kreye. 1980. *The New International Division of Labour: Structural Unemployment in Industrialized Countries and Industrialization in Developing Countries*. Cambridge: Cambridge University Press

Fuchs, R.J., and E.M. Pernia. 1987. 'External economic forces and national spatial development: Japanese direct investment in Pacific Asia.' In *Urbanization and Urban Policies in Pacific Asia*, ed. R.J. Fuchs, 88-114. Boulder: Westview Press

Gauthier, H.L. 1970. 'Geography, transportation, and regional development.' *Economic Geography* 46 (3):612-19

Gertler, M.S. 1992. 'Flexibility revisited: Districts, nation-state, and the forces of production.' *Transactions of the Institute of British Geographers* 17:259-78

Ginsburg, N. 1961. 'The dispersed metropolis: The case of Okayama.' *Toshi Mondai* 52 (6):631-40

–. 1988. 'Extended metropolitan regions in Asia: A new spatial paradigm.' Paper presented at the Chinese University of Hong Kong

–. 1990. *The Urban Transition: Reflections on the American and Asian Experiences*. Hong

Kong: Hong Kong University Press
–, B. Koppel, and T.G. McGee, eds. 1991. *Extended Metropolis: Settlement Transition in Asia*. Honolulu: University of Hawaii Press
Giuliano, G. 1986. 'Land use impacts of transportation investments: Highway and transit.' In *Geography of Urban Transportation*, ed. S. Hanson, 247-79. New York: Guilford Press
Goodman, D.S.G., ed. 1989. *China's Regional Development*. London: Routledge
–. 1994. 'Introduction: The political economy of change.' In *China's Quiet Revolution*, ed. D. Goodman and B. Hooper, ix-xxi. Melbourne: Longman Cheshire
–, and B. Hooper, eds. 1994. *China's Quiet Revolution*. Melbourne: Longman Cheshire
–, and G. Segal, eds. 1994. *China Deconstructs: Politics, Trade and Regionalism*. London: Routledge
Gottmann, J. 1961. *Megalopolis: The Urbanized Northeastern Seaboard of the United States*. Norwood, MA: Plimpton Press
Guldin, G. 1992. 'Towards a greater Guangzhou: Hong Kong's sociocultural impact on the Pearl River Delta and beyond.' Paper presented at the Workshop on Hong Kong-Guangdong Integration, University of British Columbia, Vancouver
Hansen, N.M. 1981. 'Development from above: The centre-down development paradigm.' In *Development from Above or Below?* ed. W.B. Stohr and D.R.F. Taylor, 15-38. New York: John Wiley
Harvey, D. 1989. *The Condition of Postmodernity: An Enquiry into the Origins of Cultural Change*. Oxford: Blackwell
Henderson, J., and R.P. Appelbaum. 1992. 'Situating the state in the East Asian development process.' In *States and Development in the Asian Pacific Rim*, ed. J. Henderson and R.P. Appelbaum, 1-26. Newbury Park, CA: Sage
Hettne, B. 1983. 'The development of development theory.' *Sociologica* 26 (3/4):247-66
–. 1990. *Development Theory and the Three Worlds*. London: Longman
Hirschman, A.O. 1958. *The Strategy of Economic Development*. New Haven, CT: Yale University Press
Ho, S.P.S. 1994. *Rural China in Transition*. Oxford: Clarendon Press
–, and R.W. Huenemann. 1984. *China's Open Door Policy*. Vancouver: UBC Press
Hong Kong, Federation of Hong Kong Industries. 1992. *Hong Kong's Industrial Investment in the Pearl River Delta*. Hong Kong: Federation of Hong Kong Industries
Hoyle, B.S., and D. Hilling, eds. 1984. *Seaport Systems and Spatial Change*. New York: John Wiley
Hsu, R.C. 1991. *Economic Theories in China: 1979-1988*. Cambridge: Cambridge University Press
Huang, P.C.C. 1990. *The Peasant Family and Rural Development in the Yangzi Delta: 1350-1988*. Palo Alto, CA: Stanford University Press
Hymer, S. 1972. 'The multinational corporation and the law of uneven development.' In *International Firms and Modern Imperialism*, ed. H. Radice, 37-62. London: Penguin
International Monetary Fund. 1991. *International Financial Statistics*. 44 (3-4)
Janelle, D.G. 1969. 'Spatial reorganization: A model and concept.' *Annals of the Association of American Geographers* 59 (2):348-64
–. 1986. 'Metropolitan expansion and the communications-transportation trade-off.' In *The Geography of Urban Transportation*, ed. S. Hanson, 357-85. New York: Guilford Press
Jia, H., and Z. Lin, eds. 1994. *Changing Central-Local Relations in China*. Boulder, CO: Westview Press
Johnson, C. 1982. *MITI and the Japanese Economic Miracle: The Growth of Industry Policy: 1925-1975*. Stanford: Stanford University Press
Johnson, G.E. 1979. 'Agriculture the base: Rural development 1949-79.' *International Journal* 34 (4):606-23
–. 1982. 'The production responsibility system in Chinese agriculture: Some examples from Guangdong.' *Pacific Affairs* 55 (3):430-51
–. 1986a. 'Responsibility and reform: Consequences of recent policy changes in rural South China.' *Contemporary Marxism* 12-13:144-62

–. 1986b. '1997 and after: Will Hong Kong survive? A personal view.' *Pacific Affairs* 59 (2):237-54

–. 1989. 'Rural transformation in South China: Views from the locality.' *Revue Européenne des Sciences Sociales* 84:191-226

–. 1992. 'The political economy of Chinese urbanization: Guangdong and the Pearl River Delta region.' In *Urbanizing China,* ed. G. Guldin, 185-220. Westport, CT: Greenwood Press

Keeble, D. 1967. 'Models of economic development.' In *Models in Geography,* ed. R. Chorley and P. Haggett, 287-302. London: Methuen

Keidel, A. 1991. 'The cyclical future of China's economy.' In *China's Economic Dilemmas in the 1990s,* ed. Joint Economic Committee, Congress of the United States, 119-34. Washington, DC: US Government Printing Office

Kelley, A.C., and C. Williamson. 1984. 'Population growth, industrial revolution, and the urban transition.' *Population and Development Review* 10 (3):419-41

Kirkby, R.J.R. 1985. *Urbanization in China.* New York: Columbia University Press

–. 1994. 'Dilemmas of urbanization: Review and prospects.' In *China: The Next Decades,* ed. D. Dwyer, 128-55. New York: Longman

–, and T. Cannon. 1989. 'Introduction.' In *China's Regional Development,* ed. D.S.G. Goodman, 1-19. London: Routledge

Kornai, J. 1992. *The Socialist System.* Princeton: Princeton University Press

Kumar, A. 1994. 'Economic reform and the internal division of labour in China: Production, trade and marketing.' In *China Deconstructs: Politics, Trade and Regionalism,* ed. D. Goodman and G. Segal, 99-130. London: Routledge

Kwan, C.H. 1994. *Economic Interdependence in the Asia-Pacific Region.* London: Routledge

Kwok, R.Y.W., W.L. Parish, A.G.O. Yeh, and X.Q. Xu, eds. 1990. *Chinese Urban Reform: What Model Now?* Armonk, NY: M.E. Sharpe

–, and A.Y. So, eds. 1995. *The Hong Kong-Guangdong Link.* New York: M.E. Sharpe

Lardy, N.R. 1978. *Economic Growth and Distribution in China.* Cambridge: Cambridge University Press

–. 1987. *China's Entry into the World Economy.* Lanham, MD: University Press of America

Lary, D. 1975. *Region and Nation.* Cambridge: Cambridge University Press

–. 1996. 'The tomb of the King of Nanyue: The contemporary agenda of history.' *Modern China* 22 (1):3-27

Lee, Y.S.F. 1989. 'Small towns and China's urbanization level.' *China Quarterly* 120:771-85

Leinbach, T.R., and L.S. Chia, eds. 1989. *South-East Asian Transport: Issues in Development.* Singapore: Oxford University Press

Leung, Chi-Keung. 1980. *China's Railway Patterns and National Goals.* Research Paper no. 195. Chicago: Department of Geography, University of Chicago

Leung, Chi-Kin. 1990. 'Locational characteristics of foreign equity joint venture investment in China, 1979-1985.' *Professional Geographer* 42 (4):403-21

–. 1993. 'Personal contacts, subcontracting linkages, and development in the Hong Kong-Zhujiang Delta region.' *Annals of the Association of American Geographers* 83 (2):272-302

–. 1996. 'Foreign manufacturing investment and regional industrial growth in Guangdong Province, China.' *Environment and Planning* 28:513-36

Li, S.M. 1989. 'Labour mobility, migration and urbanization in the Pearl River Delta area.' *Asian Geographer* 8 (1/2):35-60

Li, Z. 1994. *The Private Life of Chairman Mao.* New York: Random House

Lin, G.C.S. 1993. 'Small town development in socialist China: A functional analysis.' *Geoforum* 24 (3):327-38

–. 1994a. 'Changing theoretical perspectives on urbanization in Asian developing countries.' *Third World Planning Review* 16 (1):1-23

–. 1994b. 'State-led dependent development in the Asian NICs: A case study of Taiwan.' *Western Geography* 4:5-28

–, and L.J.C. Ma. 1994. 'The role of small towns in Chinese regional development.'

International Regional Science Review 17 (1):75-97

Lin, N. 1995. 'Local market socialism: Local corporatism in action in rural China.' *Theory and Society* 24:301-54

Linge, G.J.R., and D.K. Forbes, eds. 1990. *China's Spatial Economy.* Hong Kong: Oxford University Press

Lipietz, A. 1987. 'The globalization of the general crisis of Fordism: 1967-1984.' In *Frontyard/Backyard: The Americas in the Global Crisis,* ed. J. Holmes and C. Leys, 23-55. Toronto: Between the Lines

Liu, Y. L. 1992. 'Reform from below: The private economy and local politics in rural industrialization.' *China Quarterly* 130:293-316

Lo, C.P. 1987. 'Socialist ideology and urban strategies in China.' *Urban Geography* 8 (3):440-58

-. 1989. 'Recent spatial restructuring in Zhujiang Delta, South China: A study of socialist regional development strategy.' *Annals of the Association of American Geographers* 79 (2):293-308

-. 1990. 'The geography of rural regional inequality in Mainland China.' *Transactions of the Institute of British Geographers* 15:466-86

Lo, F.C., and K. Salih. 1981. 'Growth poles, agropolitan development, and polarization reversal: The debate and search for alternatives.' In *Development from Above or Below?* ed. W.B. Stohr and D.R.F. Taylor, 123-52. New York: John Wiley & Sons

Lyons, T.P., and V. Nee, eds. 1994. *The Economic Transformation of South China.* Ithaca, NY: East Asia Program, Cornell University

Ma, L.J.C. 1980. *Cities and City Planning in the People's Republic of China: An Annotated Bibliography.* HUD User Bibliography Series. US Department of Housing and Urban Development, Office of Policy Development and Research and Office of International Affairs. Washington, DC: US Government Printing Office

-, and G.H. Cui. 1987. 'Administrative changes and urban population in China.' *Annals of the Association of American Geographers* 77 (3):373-95

-, and M. Fan. 1994. 'Urbanization from below: The growth of towns in Jiangsu, China.' *Urban Studies* 31 (10):1625-45

-, and E.W. Hanten, eds. 1981. *Urban Development in Modern China.* Boulder, CO: Westview Press

-, and G.C.S. Lin. 1993. 'Development of towns in China: A case study of Guangdong Province.' *Population and Development Review* 19 (3):583-606

-, and A.G. Noble. 1986. 'Chinese cities: A research agenda.' *Urban Geography* 7 (4):279-90

Mao, Z. 1937. 'On Contradiction.' English transl. 1990. In *Mao Zedong on Dialectical Materialism,* ed. N. Knight, 154-67. Armonk, NY: M.E. Sharpe

Massey, D.B. 1984. *Spatial Divisions of Labour: Social Structures and the Geography of Production.* London: Macmillan

McGee, T.G. 1989. 'Urbanisasi or Kotadesasi? Evolving patterns of urbanization in Asia.' In *Urbanization in Asia,* ed. F.J. Costa, A.K. Dutt, L.J.C Ma, and A.G. Noble, 93-108. Honolulu: University of Hawaii Press

-. 1991a. 'Presidential address. Eurocentrism in geography: The case of Asian urbanization.' *Canadian Geographer* 35 (4):332-44

-. 1991b. 'The emergence of desakota regions in Asia: Expanding a hypothesis.' In *The Extended Metropolis: Settlement Transition in Asia,* ed. N. Ginsburg, B. Koppel, and T.G. McGee, 3-25. Honolulu: University of Hawaii Press

-, and G.C.S. Lin. 1993. 'Footprints in space: Spatial restructuring in the East Asian NICs, 1950-90.' In *Economic and Social Development in Pacific Asia,* ed. E. Dixon and D. Drakakis-Smith, 128-51. New York: Routledge

-, and I.M. Robinson, eds. 1995. *The Mega-Urban Regions of Southeast Asia.* Vancouver: UBC Press

Murphey, R. 1976. 'Chinese urbanization under Mao.' In *Urbanization and Counter-Urbanization,* ed. B.J.L. Berry, 311-29. Beverly Hills: Sage

Myrdal, G. 1957. *Economic Theory and Underdeveloped Regions.* London: Duckworth

Naughton, B. 1988. 'The third front: Defence industrialization in the Chinese interior.' *China Quarterly* 115:351-86

–. 1995. *Growing Out of the Plan.* Cambridge: Cambridge University Press

Nee, V. 1989. 'A theory of market transition: From redistribution to markets in state socialism.' *American Sociological Review* 54:663-81

Nemeth, R.J., and D.A. Smith. 1983. 'Divergent patterns of urbanization in the Philippines and South Korea: A historical structural approach.' *Comparative Urban Research* 10 (1):21-45

Nolan, P. 1994. 'Introduction: The Chinese puzzle.' In *China's Economic Reforms,* ed. Q. Fan and P. Nolan, 1-20. New York: St. Martin's Press

Ohmae, K. 1993. 'The rise of the region state.' *Foreign Affairs* 72 (2):78-87

Oi, J. 1989. *State and Peasant in Contemporary China.* Berkeley: University of California Press

–. 1995. 'The role of the local state in China's transitional economy.' *China Quarterly* 144:1132-49

Orleans, L.A. 1991. 'Loss and misuse of China's cultivated land.' In *China's Economic Dilemmas in the 1990s,* ed. Joint Economic Committee, Congress of the United States, 403-17. Armonk, NY: M.E. Sharpe

Owen, W. 1964. *Strategy for Mobility.* Washington, DC: Brookings Institution Transport Research Program

–. 1987. *Transportation and World Development.* Baltimore: Johns Hopkins University Press

Pannell, C.W. 1985. 'Recent Chinese agriculture.' *Geographical Review* 75 (2):170-85

–. 1988. 'Economic reforms and readjustment in the People's Republic of China and some geographical consequences.' *Studies in Comparative International Development* 22:54-73

–. 1989. 'Employment structure and the Chinese urban economy.' In *Urbanization in Asia,* ed. F.J. Costa, A.K. Dutt, L.J.C Ma, and A.G. Noble,, 285-308. Honolulu: University of Hawaii Press

–. 1990. 'China's urban geography.' *Progress in Human Geography* 14 (2):214-36

–, and L.J.C. Ma. 1983. *China: The Geography of Development and Modernization.* London: Edward Arnold

–, and G. Veeck. 1989. 'Zhujiang Delta and Sunnan: A comparative analysis of regional urban systems and their development.' In *The Environment and Space Development in the Pearl River Delta,* ed. X.G. Xu, A.G.O. Yeh, and C.E. Wen, 237-52. Beijing: Academic Press

–, and G. Veeck. 1991. 'China's urbanization in an Asian context: Forces for metropolitanization.' In *Extended Metropolis,* ed. N. Ginsburg et al., 113-35. Honolulu: University of Hawaii Press

Parish, W.L. 1990. 'What model now?' In *China's Urban Reform: What Model Now?* ed. R.Y.W. Kwok, W.L. Parish, A.G.O. Yeh, and X.Q. Xu, 3-5. Armonk, NY: M.E. Sharpe

Perry, E.J., and C. Wong. 1985. *The Political Economy of Reform in Post-Mao China.* Cambridge: Harvard University Press

Pieke, F.N. 1995. 'Bureaucracy, friends, and money: The growth of capital socialism in China.' *Comparative Studies in Society and History* 37 (3):494-518

Redding, S.G. 1990. *The Spirit of Chinese Capitalism.* Berlin: W. de Gruyter

Rimmer, P.J. 1995. 'Moving goods, people, and information: Putting the ASEAN mega-urban region in context.' In *Mega-Urban Regions of Southeast Asia,* ed. T.G. McGee and I.M. Robinson, 150-75. Vancouver: UBC Press

Riskin, C. 1987. *China's Political Economy: The Quest for Development Since 1949.* New York: Oxford University Press

Schuurman, F.J., ed. 1993. *Beyond the Impasse: New Directions in Development Theory.* London: Zed Books

Scott, A.J. 1988. 'Flexible production systems and regional development: The rise of new industrial spaces in North America and Western Europe.' *International Journal of Urban and Regional Research* 12 (2):171-86

–. 1992. 'The Roepke Lecture in Economic Geography. The collective order of flexible production agglomerations: Lessons for local economic development policy and strategic choice.' *Economic Geography* 68 (3):219-33

–, and M. Storper. 1992. 'Regional development reconsidered.' In *Regional Development and Contemporary Industrial Response,* ed. H. Ernste and V. Meier, 3-24. London: Belhaven Press

Seers, D. 1972. 'What are we trying to measure?' In *Measuring Development,* ed. N. Baster, 21-36. London: Frank Cass

–. 1977. 'The new meaning of development.' *International Development Review* XIX (3):2-7

Shue, V. 1988. *The Reach of the State.* Stanford: Stanford University Press

Sicular, T. 1991. 'China's agricultural policy during the reform period.' In *China's Economic Dilemmas in the 1990s,* ed., Joint Economic Committee, Congress of the United States, 340-64. Armonk, NY: M.E. Sharpe

–. 1995. 'Redefining state, plan and market: China's reforms in agricultural commerce.' *China Quarterly* 144:1020-46

Simon, D., ed. 1990. *Third World Regional Development: A Reappraisal.* London: Paul Chapman

Sit, V.F.S., ed. 1984. *Resource and Development of the Pearl River Delta.* Hong Kong: Wide Angle Press

–. 1985. *Chinese Cities.* Oxford: Oxford University Press

–, and K. Mera. 1982. *Urbanization and National Development in Asia: A Comparative Study.* Hong Kong: Comparative Urbanization Project, University of Hong Kong

Skeldon, R., ed. 1994. *Reluctant Exiles?: Migration from Hong Kong and the New Overseas Chinese.* Hong Kong University Press

Skinner, G.W. 1977. 'Regional urbanization in Nineteenth Century China.' In *The City in Late Imperial China,* ed. G.W. Skinner, 211-29. Stanford: Stanford University Press

–. 1985. 'Rural marketing in China: Repression and revival.' *China Quarterly* 103:393-413

–. 1994. 'Differential development in Lingnan.' In *The Economic Transformation of South China,* ed. T.P. Lyons and V. Nee, 17-54. Ithaca, NY: East Asia Program, Cornell University

–, and E.A. Winckler. 1969. 'Compliance succession in rural Communist China: A cyclical theory.' In *A Sociological Reader on Complex Organization,* ed. A. Etzioni, 410-38. New York: Holt, Rinehart and Winston

Smart, A. 1993. 'Gifts, bribes, and Guanxi: A reconsideration of Bourdieu's social capital.' *Cultural Anthropology* 8 (3):388-408

–. 1995. 'Local capitalisms: situated social support for capitalist production in China.' Occasional Paper no. 121, Department of Geography, Chinese University of Hong Kong

–, and J. Smart. 1992. 'Capitalist production in a Socialist society: The transfer of manufacturing from Hong Kong to China.' In *Anthropology and the Global Factory,* ed. F. Rothstein and M. Blim, 47-61. Westport, CT: Bergin and Garvey

Smart, J., and A. Smart. 1991. 'Personal relations and divergent economies: A case study of Hong Kong investment in China.' *International Journal of Urban and Regional Research* 15 (2):216-33

Solinger, D.J. 1985. 'Marxism and the market in socialist China: The reforms of 1979-1980 in context.' In *State and Society in Contemporary China,* ed. V. Nee and D. Mozindo, 194-222. Ithaca, NY: Cornell University Press

–. 1987. 'Uncertain paternalism: Tensions in recent regional restructuring in China.' *International Regional Science Review* 11 (1):23-42

–. 1989. 'Capitalist measures with Chinese characteristics.' *Problems of Communism* 38 (1):19-33

–. 1993. *China's Transition from Socialism.* Armonk, NY: M.E. Sharpe

Stohr, W.B. 1981. 'Development from below: The bottom up and periphery inward development paradigm.' In *Development from Above or Below?* ed. W.B. Stohr and D.R.F. Taylor, 39-71. New York: John Wiley

–, and D.R.F. Taylor, ed. 1981. *Development from Above or Below?* New York: John Wiley

Storper, M., and A.J. Scott. 1989. 'The geographical foundations and social regulation of flexible production complexes.' In *The Power of Geography*, ed. J. Wolch and M. Dear, 21-40. Boston: Unwin Hyman

Sung, Y.W. 1991. *The China-Hong Kong Connection: The Key to China's Open-Door Policy*. Cambridge: Cambridge University Press

–, P.W. Liu, Y.C. Wong, and P.K. Lau. 1995. *The Fifth Dragon: The Emergence of the Pearl River Delta*. Singapore: Addison Wesley

Taaffe, E.J., and H.L. Gauthier. 1973. *Geography of Transportation*. Englewood Cliffs, NJ: Prentice-Hall

Tan, K.C. 1986a. 'Revitalized small towns in China.' *Geographical Review* 76 (2):138-48

–. 1986b. 'Small towns in Chinese urbanization.' *Geographical Review* 76 (3):265-75

Tang, W.S. 1994. 'Urban land development under socialism: China between 1949 and 1977.' *International Journal of Urban and Regional Research* 18 (3):392-415

Taylor, B., and R.Y.W. Kwok. 1989. 'From export center to world city: Planning for the transformation of Hong Kong.' *Journal of the American Planning Association* 55:309-22

Thant, M., M. Tang, and H. Kakazu, eds. 1994. *Growth Triangles in Asia*. Hong Kong: Oxford University Press

Theroux, P. 1993. 'Going to see the dragon.' *Harper's Folio*, October, 33-56

Thoburn, J.T., H.M. Leung, E. Chau, and S.H. Tang. 1990. *Foreign Investment in China under the Open Policy*. Hong Kong: Gower Publishing

Veeck, G., and C.W. Pannell. 1989. 'Rural economic restructuring and farm household income in Jiangsu, People's Republic of China.' *Annals of the Association of American Geographers* 79 (2):275-92

Vogel, E. 1989. *One Step Ahead in China: Guangdong under Reform*. Cambridge, MA: Harvard University Press

Vohra, R. 1994. 'Deng Xiaoping's modernization: Capitalism with Chinese characteristics.' *Developing Societies* 10:46-58

Walder, A.G. 1995a. 'Local governments as industrial firms: An organizational analysis of China's transitional economy.' *American Journal of Sociology* 101 (2):263-301

–. 1995b. 'China's transitional economy: Interpreting its significance.' *China Quarterly* 144:963-79

–. 1996. 'Markets and inequality in transitional economies: Toward testable theories.' *American Journal of Sociology* 101 (4):1060-73

Walker, K.R. 1984. 'Chinese agriculture during the period of the readjustment, 1978-1983.' *China Quarterly* 100:783-812

Wank, D. 1993. 'From state socialism to community capitalism: State power, social structure, and private enterprise in a Chinese city.' PhD dissertation. Harvard University

Webb, S.E.H., and F.C. Tuan. 1991. 'China's agricultural reform: Evaluation and outlook.' In *China's Economic Dilemma in the 1990s*, ed. Joint Economic Committee, Congress of the United States, 340-64. Armonk, NY: M.E. Sharpe

Wei, Y., and L.J.C. Ma. 1996. 'Changing patterns of spatial inequality in China, 1952-1990.' *Third World Planning Review* 18 (2):177-91

Whebell, C.F.J. 1969. 'Corridors: A theory of urban system.' *Annals of the Association of American Geographers* 59 (1):1-26

Wilson, G.W., ed. 1966. *The Impact of Highway Investment on Development*. Washington, DC: Brooking Institute Transport Research Program

Wong, K.Y., and S. Tong. 1984. 'Geography of the Pearl River Delta: An introduction.' In *Resource and Development of the Pearl River Delta*, ed. V.F. Sit, 3-23. Hong Kong: Wide Angle Press

Wu, C.T. 1987. 'Chinese socialism and uneven development.' In *The Socialist Third World*, ed. D. Forbes and N. Thrift, 53-97. Oxford: Basil Blackwell

–, and D.F. Ip. 1981a. 'China: Rural development: Alternating combinations of top-down and bottom-up strategies.' In *Development from Above or Below?* ed. W.B. Stohr and D.R.F. Taylor, 155-82. New York: John Wiley

–, and D.F. Ip. 1981b. 'Structural transformation and spatial equity.' In *China: Urbanization and National Development,* ed. C.K. Leung and N. Ginsburg, 56-88. Research Paper No. 196. Chicago: University of Chicago, Department of Geography

Wu, Y.L. 1967. *The Spatial Economy of Communist China.* New York: Praeger

Xu, X.Q. 1984. 'Characteristics of urbanization in China: Changes and causes of urban population growth and distribution.' *Asian Geographer* 3 (1):15-19

–, and S.M. Li. 1990. 'China's open door policy and urbanization in the Pearl River Delta region.' *International Journal of Urban and Regional Research* 14 (1):49-69

–, N. Ouyang, and C. Zhou. 1995. 'The Changing urban system of China: New developments since 1978.' *Urban Geography* 16 (6):493-504

Yang, D. 1990. 'Patterns of Chinese regional development strategy.' *China Quarterly* 122:230-57

Ye, S.Z. 1989. 'Urban development trends in China.' In *Urbanization in Asia,* ed. F.J. Costa, A.K. Dutt, L.J.C. Ma, and A.G. Noble, 75-92. Honolulu: University of Hawaii Press

Yee, F.L. 1992. 'Economic and urban changes in the Shenzhen Special Economic Zone 1979-1986.' PhD dissertation, University of British Columbia

Yeh, A.G.O., K.C. Lam, S.M. Li, and K.Y. Wong. 1989. 'Spatial development of the Pearl River Delta: Development issues and research agenda.' *Asian Geographer* 8 (1/2):1-9

–, and X.Q. Xu. 1996. 'Globalization and the urban system in China.' In *Emerging World Cities in Pacific Asia,* ed. F.C. Lo and Y.M. Yeung, 219-67. Tokyo: United Nations University Press

Yeh, K.C. 1984. 'Macroeconomic changes in the Chinese economy during the readjustment.' *China Quarterly* 100:691-716

Yeung, Y.M. 1994. 'Infrastructure development in the Southern China growth triangle.' In *Growth Triangles in Asia,* ed. M. Thant, M. Tang, and H. Kakazu, 114-50. Hong Kong: Oxford University Press

–, and D.K.Y. Chu, eds. 1994. *Guangdong: Survey of a Province Undergoing Rapid Change.* Hong Kong: Chinese University of Hong Kong Press

–, and X. Hu, eds. 1992. *China's Coastal Cities.* Honolulu: University of Hawaii Press

–, and F. Lo. 1996. 'Global restructuring and emerging urban corridors in Pacific Asia.' In *Emerging World Cities in Pacific Asia,* ed. F.C. Lo and Y.M. Yeung, 17-47. Tokyo: United Nations University Press

Zhong, G.F. 1982. 'The mulberry dike-fish pond complex: A Chinese ecosystem of land-water interaction on the Pearl River Delta.' *Human Ecology* 10 (2):191-202

Zhou, Y.X. 1988. 'On the relationship between urbanization and gross national product.' *Chinese Sociology and Anthropology* 21 (2):3-16

–. 1991. 'The metropolitan interlocking region in China: A preliminary hypothesis.' In *Extended Metropolis,* ed. N. Ginsburg, B. Koppel, and T.G. McGee, 89-112. Honolulu: University of Hawaii Press

Publications in Chinese

Cao, Y. 1990. 'The theories of growth pole and growth centre: A summary.' *Dili Yanjiu* (Geographical Research) 9 (3):87-94

CCP Team (Chinese Communist Party, Central Committee Special Investigation Team). 1989. *Dongguan Shinian* (Dongguan's Ten Years). Shanghai: People's Publishing House of Shanghai

Chen, G.D. 1934. 'Guangzhou sanjiaozhou wenti' (On the Guangzhou Delta). *Kexue* (Science) 3

Chen, L. 1992. *Nanhaishi Tudiliyong Zhongti Guihua* (A Master Plan for Land Use for Nanhai). Internal document

China, State Statistical Bureau (SSB). 1982. *Zhongguo Tongji Nianjian* (China's Statistical Yearbook). Hong Kong: Hong Kong Jingji daobaoce

–. 1983. *Zhongguo Tongji Nianjian* (China's Statistical Yearbook). Beijing: China Statistical Press

–. 1985. *Zhongguo Chengshi Tongji Nianjian* (Statistical Yearbook for Chinese Cities). Beijing: New World Press

–. 1988. *Zhongguo 1987 Nian 1% Renkou Chouyang Diaocha Ziliao: Quanguo Fence* (Tabulations of China's 1% Population Sample Survey of 1987: National Volume). Beijing: China Statistical Press

–. 1991a. *Zhongguo Tongji Zhaiyao* (China's Statistical Abstract). Beijing: China Statistical Press

–. 1991b. *Zhongguo Chengshi Tongji Nianjian* (Statistical Yearbook for Chinese Cities). Beijing: New World Press

–. 1991c. *Zhongguo Tongji Nianjian* (China's Statistical Yearbook). Beijing: China Statistical Press

–. 1995. *Zhongguo Tongji Nianjian* (China's Statistical Yearbook). Beijing: China Statistical Press

–. 1996. *Zhongguo Tongji Zhaiyao* (China's Statistical Abstract). Beijing: China Statistical Press

Chinese Academy of Sciences, Geographic Institute. 1987. *Zhongguo Gongye Fenbu Tuji* (Atlas of Chinese Industry). Beijing: Planning Press

Chinese News Agency. 1992. *Zhongguo Xinwen* (Chinese News). Beijing: Chinese News Agency, 12632

Dalizhen. 1992a. *Nanhaixian Dalizhen Jianjie* (Information on Dali Township of Nanhai County). Internal document obtained from an interview

–. 1992b. *Nanhaixian Dalizhen Lianjiao Guanliqu Jianjie* (Information on Lianjiao District, Dali Township of Nanhai County). Internal document obtained from an interview

–. 1992c. *Nanhaixian Dalizhen Lianjiao Guanliqu Tuixing Liangshi Shengchan Qiye Jingying de Shiyan Fang'an Jiqi Shexiang Anpai* (A Proposal for the Enterprisation of Rice Cultivation in Lianjiao District). Internal document obtained from interview

Dongguan. 1992. Interviews conducted 12 September to 10 October

Dongguan, Statistical Bureau. 1991. *Dongguan Tongji Nianjian (1978-1990)* (Statistical Yearbook for Dongguan 1978-90). Dongguan: Internal document

Foshan, Statistical Bureau. 1989. *Foshanshi Tongji Ziliao Huibian (1949-1988)* (A Collection of Statistical Data for Foshan Municipality, 1949-88). Foshan: Internal document

Foshan and Nanhai, Joint Investigation Team. 1987. 'Nanhaixian Jinnianlai Nongye Jixiehua de Qingkuan Fenxi' (An analysis of recent agricultural mechanization in Nanhai). *Nanfang Nongcun* (South China Countryside) 13:35-7

Guangdong, Land Development Bureau. 1986. *Guangdongsheng Guotu Ziyuan* (Land Resources of Guangdong Province). Guangzhou: Internal document

Guangdong, Office for Population Census. 1988. *Zhongguo 1987 Nian 1% Renkou Chouyang Diaocha Ziliao: Guangdong Fence* (Tabulations of China's 1% Population Sample Survey of 1987: Guangdong Volume). Beijing: China Statistical Press

–. 1991. *Guangdongsheng Disici Renkou Pucha Shougong Huizhong Ziliao* (Manual Processed Data for the 4th Population Census for Guangdong Province). Guangzhou: Internal document

Guangdong, Statistical Bureau. 1981. *Guangdongsheng Guomin Jingji Tongji Zaiyao (1949-1980)* (Statistical Information for Guangdong Province, 1949-80). Guangzhou: Confidential document

–. 1985. *Guangdongsheng Tongji Nianjian* (Statistical Yearbook of Guangdong Province). Hong Kong: Jingji daobaoshe

–. 1986. *Guangdongsheng Xiaochengzhen Diaocha Ziliao Huibian* (A Collection of Investigation Data on the Small Towns of Guangdong Province). Guangzhou: Internal document

–. 1987. *Guangdongsheng Guding Zichan Touzi Tongji Ziliao (1986)* (Statistical Information on Fixed Assets Investment of Guangdong Province 1986). Guangzhou: Confidential document

–. 1990. *Guangdong Tongji Nianjian* (Statistical Yearbook of Guangdong). Beijing: China Statistical Press

–. 1991a. *Guangdong Tongji Nianjian* (Statistical Yearbook of Guangdong Province). Beijing: China Statistical Press

–. 1991b. *Guangdongsheng Xianqu Guomin Jingji Tongji Ziliao (1980-1990)* (National Economic Statistical Data for Cities and Counties of Guangdong Province, 1980-90). Guangzhou: Internal publication

–. 1991c. *Guangdongsheng Duiwai Jingji Maoyi Luyou Tongji Ziliao* (Statistical Information on Foreign Trade and Tourism for Guangdong Province). Guangzhou: Internal document

–. 1992a. *Guangdong Tongji Nianjian* (Statistical Yearbook of Guangdong Province). Beijing: China Statistical Press

–. 1992b. *Zhujiang Sanjiaozhou Guomin Jingji Tongji Ziliao (1980-1991)* (National Economic Statistical Data for the Zhujiang Delta, 1980-91). Guangzhou: Internal publication

Guangdong, Foshan, and Nanhai, Joint Investigation Team. 1989. 'Guanyu nongcun tudi zhidu jianshe de jige wenti' (On establishing rural land systems). *Nongcun yanjiu* (Rural Studies) 39:9-22

Guangzhou, Statistical Bureau. 1989. *Guangzhou Tongji Nianjian* (Statistical Yearbook of Guangzhou). Beijing: China Statistical Press

–. 1990. *Guangzhou Tongji Nianjian* (Statistical Yearbook of Guangzhou). Beijing: China Statistical Press

–. 1991. *Guangzhou Tongji Nianjian* (Statistical Yearbook of Guangzhou). Beijing: China Statistical Press

–. 1992. *Guangzhou Tongji Nianjian* (Statistical Yearbook of Guangzhou). Beijing: China Statistical Press

Hu, Z.L., and P.M. Foggin. 1993. 'Gaige kaifang zhengce yi chengshi fazhan' (The reform and open-door policy versus urban development). *Chengshi Kexue* (Urban Science) 3:33-7

–, and P.M. Foggin. 1994. *Gaige Kaifanghou de Zhongguo Chengshi* (Chinese Cities Since Economic Reform). IDRC project report. Beijing: Peking University

–, and X.C. Meng. 1988. 'Socio-economic and political background of the growth of large cities in China.' *Asian Geographer* 6 (1):24-8

Huang, G.Y., and P.S. Zhong. 1958. 'Guanyu zhujiang sanjiaozhou de jige wenti' (On the Zhujiang Delta). *Dilixue Ziliao* (Geographical Information) 3

Huang, H.S., ed. 1991. *Dongguan Nongye Jingji* (Dongguan's Agricultural Economy). Guangzhou: People's Publishing House of Guangdong

Jiang, C. 1990. *Lunzhuhou Jingji* (On Feudal Lord Economies). *Jingji Wenti Tansuo* (Inquiry on Economic Issues) 5:10-14

Jingji Cankao (Economy Reference)

Li, M.B., and X. Hu, eds. 1991. *Liudong Renkou Dui Dachengshi Fazhan de Yingxiang ji Duice* (The Impact of Floating Population on the Development of Large Cities and Responding Strategies). Beijing: Economy Daily Press

Li, R.Q. 1988. 'The growth pole theory in China's regional development and the study of policies.' *Jingji Yanjiu* (Economic Research) 4:63-70

Liao, J.H. 1992. 'Guangdong Nanhaishi shui huanjing zhiliang' (The environmental quality of water in Nanhai Municipality). *Ziran Dili yu Huanjing Yanjiu* (Research on Physical Geography and Environment), ed. GeoScience College, Zhongshan University, 273-81. Guangzhou: Zhongshan University Press

Lin, G.C.S. 1986. *Guangzhou Chengshi Fazhan Fenxi* (Urban Development of Guangzhou: An Analysis). Guangzhou: People's Publishing House of Guangdong

–, and L.J.C. Ma. 1990. 'Woguo xiaochengzhen gongneng jiegou chutan' (Functional structure of small towns in China). *Dili Xuebao* (Acta Geographica Sinica) 45 (4):412-20

Lu, P. 1992. 'Zhujiang sanjiaozhou-Xianggang jingji jishu hezuo de huigu yu qianzhan' (A retrospect and prospect on economic and technological cooperation between the Zhujiang Delta and Hong Kong). In *Zhujiang Sanjiaozhou Jingji Fazhan Huigu yu Qianzhan* (Economic Development in the Zhujiang Delta: Retrospect and Prospect), ed. Research Center for Economic Development and Management in the Zhujiang Delta, 141-50. Guangzhou: Zhongshan University Press

Miao, H.J., C.X. Shen, and G.Y. Huang. 1988. *Zhujiang Sanjiaozhou Shuitu Ziyuan* (Natural Resources of the Zhujiang Delta). Guangzhou: Zhongshan University Press

Nanhai. 1992. Interviews conducted 11 - 25 October

Nanhai, Agricultural Commission. 1987. *Guangdongsheng Nanhaixian 1986 Niandu Nongchanpin Chengben Diaocha Ziliao* (Investigation Materials on the Cost of Agricultural Production in Nanhai County 1986). Nanhai: Internal document

Nanhai, Bureau of Township and Village Enterprises. 1992. *Nanhaixian Xiangzhen Qiye Jiben Qingkuang* (Basic Statistical Information for the Township and Village Enterprises in Nanhai County). Nanhai: Internal document

Nanhai, Environment Monitoring Station. 1989. *Nanhaixian Huanjing Zhiliang Nianbao (1988)* (Annual Report on the Environmental Quality of Nanhai County, 1988). Nanhai: Internal document

–. 1991. *Nanhaixian Huanjing Zhiliang Nianbao* (1990) (Annual Report on the Environmental Quality of Nanhai County, 1990). Nanhai: Internal document

Nanhai, Executive Office. 1987. 'Nanhaixian jigeng fuwu de jizhong xingshi' (On the manner in which mechanical services were delivered in Nanhai County). *Nanfang Nongcun* (South China Countryside) 13:38-40

Nanhai, Population Census Bureau. 1992a. *Guangdongsheng Nanhaixian Disici Renkou Pucha Baogaoshu* (A Report for the 4th Population Census of Nanhai County). Nanhai: Internal document

–. 1992b. *Guangdongsheng Nanhaixian Disici Renkou Pucha Jiqi Huizhong Ziliao* (Machinery Processed Data for the 4th Population Census of Nanhai County). Nanhai: Internal document

Nanhai, Statistical Bureau. 1989. *Nanhai 40 Nian* (1949-1988) (Nanhai's Four Decades, 1949-88). Nanhai: Internal document

–. 1991. *Nanhaixian 1990 Nian Tongji Ziliao* (Statistical Data for Nanhai County 1990). Nanhai: Internal document

–. 1992. *Nanhaixian 1991 Nian Tongji Ziliao* (Statistical Data for Nanhai County 1991). Nanhai: Internal document

Nanhai, Work Team. 1988a. 'Fazhan chuanghui nongye, jiasu nongye xiandaihua' (Developing export agriculture and promoting agricultural mechanization). *Nongcun Yanjiu* (Rural Studies) 29:8-13

–. 1988b. 'Nanhaixian lishui shucai chukou shengchan tixi tansuo' (A study of the vegetable export production system in Lishui). *Nongcun Yanjiu* (Rural Studies) 29:14-18

–. 1988c. 'Nanhaixian cujin nongye guimo jingying de qingkuang he mianlin de wenti' (Progress and problems in the promotion of agricultural production at a sizeable scale in Nanhai County). *Nongcun Yanjiu* (Rural Studies) 30/31:76-82

Pan, C. 1991. 'Quyu fazhan yanjiu–Zhongguo dilixue de xinmailuo' (The study of regional development–New directions of Chinese Geography). *Jingji Dili* (Economic Geography) 11 (4):1-6

Pan, L.X. et al, H.B. Cao, and Y.Z. Yu. 1991. *Guangdong Zhengqu Yanbian* (The Evolution of Administrative Districts in Guangdong). Guangzhou: Cartographic Publishing House of Guangdong

Panyu. 1992. Interviews conducted 29 August - 11 September

Panyu, Land Development Section. 1992. *Panyuxian Chengzhen Yongti Nianbao* (A Report of Land Use Change in Panyu). Panyu: Internal document

Panyu, Office of Place Record. 1989. *Panyu Dimingzhi* (Place Records for Panyu). Guangzhou: Map Publisher of Guangdong Province

Panyu, Population Census Office. 1991. *Guangzhoushi Panyuxian 1990 Nian Renkou Pucha Ziliao* (Population Census Data for Panyu, 1990). Panyu: Internal document

Panyu, Statistical Bureau. 1983. *Panyuxian Guomin Jingji Tongji Ziliao (1949-1982)* (Statistical Information for the Economy of Panyu, 1949-82). Panyu: Internal document

–. 1989. *Panyuxian Gege Shiqi Tongji Ziliao (1949-1988)* (Statistical Information for the Economy of Panyu for Major Historical Periods, 1949-88). Panyu: Internal document

–. 1991. *Panyuxian 1990 Nian Guomin Jingji Tongji Ziliao* (Statistical Information for the Economy of Panyu 1990). Panyu: Internal document

–. 1992. *Panyuxian 1991 Nian Guomin Jingji Tongji Ziliao* (Statistical Information for the Economy of Panyu 1991). Panyu: Internal document

Panyu, Urban Planning Section. 1992. *Panyuxian Chengzhen Zhongti Guihua Shuomingshu* (A Report of the Master Plan for Panyu). Panyu: Internal document

Panyubao (Panyu Daily)

Renmin Ribao (People's Daily – Overseas Edition)

Shen, L.R., and Y.C. Dai. 1990. 'Woguo zhuhou jingji de xingcheng jichi biduan hegenyuan' (Formation, problems, and origins of feudal lord economies in our country). *Jingji Yanjiu* (Economic Research) 3:12-19

Song, J.T., and G.L. Gu. 1988. 'Chengzhen tixi guihua de lilun yu fangfa chutan' (Theory and methodology of urban system planning). *Dili Xuebao* (Acta Geographica Sinica) 43:97-107

Xie, S.J., and J. Li. 1992. 'Nanhaishi jingji fazhan dui daqi huanjing yingxiang fenxi' (The impact of economic development on the atmospheric environment of Nanhai City). *Ziran Dili yu Huanjing Yanjiu* (Research on Physical Geography and Environment), ed. GeoScience College, Zhongshan University, 268-73. Guangzhou: Zhongshan University Press

Xu, C.M. 1973. *Zhujiang Sanjiaozhou* (The Zhujiang Delta). Guangzhou: People's Publishing House of Guangdong

Xu, X.Q. 1992. 'Zhujiang sanjiaozhou chengshihua de huiqu yu qianzhan' (Urbanization of the Zhujiang Delta: Retrospect and prospect). *Zhujiang Sanjiaozhou Jingji Fazhan Huigu yu Qiangzhan* (Economic Development of Zhujiang Delta: Retrospect and Prospect), ed. Research Center of Economic Development and Management for Zhujiang Delta, 13-21. Guangzhou: Zhongshan University Press

–, Q. Liu, and X.Z. Zeng, eds. 1986. 'Duiwai kaifang zhengce dui Zhujiang sanjiaozhou chengshihua jincheng de yingxiang' (The impact of the open door policy on the urbanization process in Zhujiang Delta). Paper presented at Zhongshan University, Guangzhou

–, Q. Liu, and X. Z. Zeng, eds. 1988. *Zhujiang Sanjiaozhou de Jingji Fazhan yu Chengshihua* (Development and Urbanization of Zhujiang Delta). Guangzhou: Zhongshan University Press

Wang, G.Z., B.S. Zhang, D.Z. Zhao, Z. Zhuo, and B. Liu. 1991. *Guangdong Sixiaohu* (The Four Little Tigers in Guangdong). Guangzhou: Higher Education Press of Guangdong Province

Wu, S.S., and Z.S. Zeng. 1947. 'Zhujiang sanjiaozhou' (The Zhujiang Delta). *Ningnan Xuebao* (South China Journal) 1

Yang, J.C. 1992. *Wuge Lunzi Yiqizhuan-Nanhaixian Fazhan Xiangzhen Qiye Shijian Moshi* (Driving Forward on Five Wheels: Experience of Township and Village Enterprises Development in Nanhai County). Nanhai: internal document

Yangcheng Wangbao (Guangzhou Evening Post)

Yao, S.M., and C.C. Wu. 1982. 'Woguo nongcun renkou chengshihua de yizhong teshu xingshi' (A special form of urbanization of rural population in China). *Dili* Xuebao (Acta Geographica Sinica) 37:155-63

Yatai Jingji Shibao (Asia-Pacific Economy Daily)

Yie, N., and D. Tan. 1990. 'Quyu fazhan yanjiu de lilun jincheng' (Theoretical progress of the study of regional development). *Jingji Dili* (Economic Geography) 10 (4):1-6

Zeng, G.S. 1988. 'Zhujiang sanjiaozhou jingji tizhi gaige shinian huigu' (A retrospect of economic reforms of the Pearl River Delta). *Shijian de Guanghui* (The Radiance of Practice), ed. Editorial Board of *Guangzhou Daily*, 1-7. Guangzhou: Advanced Education Press of Guangdong

Zhang, C.U. 1985. 'Nanhaixian tiaozheng nongye chanye jiegou chuyi' (On the agricultural restructuring of Nanhai County). *Nongcun Yanjiu* (Rural Studies) 14 (5):22-26

Zhao, D.Z. 1988. 'Wuge lunzi yiqizhuan' (Driving forward on five wheels). *Shijian de Guanghui* (The Radiance of Practice), ed. Editorial Board of *Guangzhou Daily*, 351-9. Guangzhou: Advanced Education Press of Guangdong

–. 1991. 'Nanhai moshi' (The Nanhai model). *Guangdong sixiaohu* (The Four Little Tigers in Guangdong), ed. G.Z. Wang, B.S. Zhang, D.Z. Zhao, Z. Zhuo, and B. Liu, 138-96. Guangzhou: Advanced Education Press of Guangdong

Zhao, S.J., G.M. Lu, and J.H. Li. 1988. 'Fada diqu tudi xiang zhongtian nengshou jizhong de jizhong xingshi' (On the manners in which cultivated land was concentrated to the rice-growing specialists in economically advanced areas). *Nanfang Nongcun* (South China Countryside) 14 (2):31-2

Zheng, T.X. 1991. *Zhujiang Sanjiaozhou Jingji Dili Wangluo* (Economic and Geographic Network of the Zhujiang Delta). Guangzhou: Zhongshan University Press

Zhong, G.F., and C.M. Li. 1960. *Zhujiang Sanjiaozhou* (The Zhujiang Delta). Shanghai: Commercial Press

Zhou, Y. 1995. *Chengshi Dilixue* (Urban Geography). Beijing: Commercial Press

Index

Set in Stone by First Folio Resource Group, Inc.

Printed and bound in Canada by Friesens

Copy-editor: Maureen Nicholson

Proofreader: Mary Williams

Indexer: Carol Graham and Margalo Whyte